T0324755

Springer Theses

Recognizing Outstanding Ph.D. Research

Aims and Scope

The series "Springer Theses" brings together a selection of the very best Ph.D. theses from around the world and across the physical sciences. Nominated and endorsed by two recognized specialists, each published volume has been selected for its scientific excellence and the high impact of its contents for the pertinent field of research. For greater accessibility to non-specialists, the published versions include an extended introduction, as well as a foreword by the student's supervisor explaining the special relevance of the work for the field. As a whole, the series will provide a valuable resource both for newcomers to the research fields described, and for other scientists seeking detailed background information on special questions. Finally, it provides an accredited documentation of the valuable contributions made by today's younger generation of scientists.

Theses are accepted into the series by invited nomination only and must fulfill all of the following criteria

- They must be written in good English.
- The topic should fall within the confines of Chemistry, Physics, Earth Sciences, Engineering and related interdisciplinary fields such as Materials, Nanoscience, Chemical Engineering, Complex Systems and Biophysics.
- The work reported in the thesis must represent a significant scientific advance.
- If the thesis includes previously published material, permission to reproduce this must be gained from the respective copyright holder.
- They must have been examined and passed during the 12 months prior to nomination.
- Each thesis should include a foreword by the supervisor outlining the significance of its content.
- The theses should have a clearly defined structure including an introduction accessible to scientists not expert in that particular field.

More information about this series at http://www.springer.com/series/8790

Mizuho Yabushita

A Study on Catalytic Conversion of Non-Food Biomass into Chemicals

Fusion of Chemical Sciences and Engineering

Doctoral Thesis accepted by
Hokkaido University, Sapporo, Japan

 Springer

Author
Dr. Mizuho Yabushita
Graduate School of Chemical Sciences and
   Engineering
Hokkaido University
Sapporo
Japan

Supervisor
Prof. Atsushi Fukuoka
Institute for Catalysis
Hokkaido University
Sapporo
Japan

ISSN 2190-5053   ISSN 2190-5061   (electronic)
Springer Theses
ISBN 978-981-10-0331-8   ISBN 978-981-10-0332-5   (eBook)
DOI 10.1007/978-981-10-0332-5

Library of Congress Control Number: 2015960223

This Springer imprint is published by SpringerNature
The registered company is Springer Science+Business Media Singapore Pte Ltd.

Parts of this thesis have been published in the following journal articles and book:

1. Kobayashi H, Yabushita M, Komanoya T, Hara K, Fujita I, Fukuoka A (2013) High-Yielding One-Pot Synthesis of Glucose from Cellulose Using Simple Activated Carbons and Trace Hydrochloric Acid. ACS Catal 3(4):581–587
2. Yabushita M, Kobayashi H, Fukuoka A (2014) Catalytic transformation of cellulose into platform chemicals. Appl Catal B Environ 145:1–9
3. Yabushita M, Kobayashi H, Hasegawa J, Hara K, Fukuoka A (2014) Entropically Favored Adsorption of Cellulosic Molecules onto Carbon Materials through Hydrophobic Functionalities. ChemSusChem 7(5):1443–1450
4. Yabushita M, Kobayashi H, Hara K, Fukuoka A (2014) Quantitative evaluation of ball-milling effect on hydrolysis of cellulose catalysed by activated carbons. Catal Sci Technol 4(8):2312–2317
5. Yabushita M, Kobayashi H, Shrotri A, Hara K, Ito S, Fukuoka A (2015) Sulfuric Acid-Catalyzed Dehydration of Sorbitol: Mechanistic Study on Preferential Formation of 1,4-Sorbitan. Bull Chem Soc Jpn 88(7):996–1002
6. Kobayashi H, Yabushita M, Hasegawa J, Fukuoka A (2015) Synergy of Vicinal Oxygenated Groups of Catalysts for Hydrolysis of Cellulosic Molecules. J Phys Chem C 119(36):20993–20999
7. Yabushita M, Kobayashi H, Kuroki K, Ito S, Fukuoka A (2015) Catalytic Depolymerization of Chitin with Retention of N-Acetyl Group. ChemSusChem 8(22):3760–3763
8. Kobayashi H, Yabushita M, Fukuoka A (2016) Depolymerization of Cellulosic Biomass Catalyzed by Activated Carbons. In: Schlaf M, Zhang ZC (eds) Reaction Pathways and Mechanisms in Thermocatalytic Biomass Conversion I: Cellulose Structure, Depolymerization and Conversion by Heterogeneous Catalysts. Springer Singapore, Singapore, pp 15–26

Supervisor's Foreword

Reduction of greenhouse gas emissions is a current global issue. To solve this problem, considerable attention has been focused on a biorefinery that can produce fuels and chemicals from renewable biomass. One of the key issues in that biorefinery is the utilization of non-food biomass, and lignocellulose is the most abundant non-food biomass on earth. However, the production of fuels and chemicals from lignocellulose has been a challenge owing to its recalcitrant structures consisting of cellulose, hemicellulose, and lignin. A similar situation can be seen for marine biomass; chitin is the most abundant marine biomass, but the selective conversion of chitin into chemicals has been a target. Enzymes and chemical catalysts have been used for the depolymerization of lignocellulose and chitin into monomers of their components, but known processes have difficulties in terms of cost, activity, separation, and durability of the enzymes and catalysts.

In this thesis, Dr. Mizuho Yabushita studied the conversion of non-food biomass into platform chemicals, mainly by solid catalysts. First, he focused on the hydrolysis of cellulose and succeeded in high-yield synthesis of glucose by weakly acidic carbon catalysts. Then he attained the production of glucose and xylose from bagasse pulp as a real biomass using carbon catalysts. Mix-milling of biomass substrates and solid catalysts greatly promotes the hydrolysis of polysaccharides. Also very important is his detailed study of the mechanism for hydrolysis of cellulose. He found that larger sugar molecules are favorably adsorbed on the carbon surface via hydrophobic interactions, and then the adsorbed molecules are hydrolyzed by the weakly acidic sites on the carbon. This mechanism resembles that of enzymes, but he showed the advantage of solid carbon catalysts that can work under a wide range of reaction conditions. Reduction of glucose gives sorbitol, and Dr. Yabushita realized the selective dehydration of sorbitol to 1,4-sorbitan as a valuable compound. He also achieved the catalytic depolymerization of chitin to monomers with retention of N-acetyl groups.

His findings will contribute to establishing new processes of catalytic conversion of non-food biomass in biorefineries and, moreover, to implementing a sustainable society in the future.

Sapporo, Japan   Prof. Atsushi Fukuoka
November 2015

Acknowledgements

This doctoral thesis work, *A Study on Catalytic Conversion of Non-Food Biomass into Chemicals: Fusion of Chemical Sciences and Engineering*, supervised by Prof. Atsushi Fukuoka [Institute for Catalysis (formerly called Catalysis Research Center until the end of September 2015), Hokkaido University] was conducted from April 2010 to March 2015. The original version of this thesis was accepted by the Graduate School of Chemical Sciences and Engineering, Hokkaido University in March 2015. This book is the revised version including additional results and discussion.

First of all, I would like to show my greatest appreciation to my supervisor, Prof. Fukuoka. His wide knowledge, invaluable suggestions, and warm encouragement supported and motivated me at all times.

I owe my deep gratitude to all the collaborators: Mr. Ichiro Fujita (Showa Denko K.K.) for supplying alkali-activated carbons and bagasse kraft pulp; Prof. Jun-ya Hasegawa (Hokkaido University) for DFT calculations; Dr. Yukiyasu Yamakoshi (Hokkaido Research Organization Industrial Research Institute) for LC/MS analyses; Prof. Wataru Ueda and Dr. Satoshi Ishikawa (Kanagawa University) for micropore measurements; Dr. Yasuhito Koyama (Hokkaido University) for discussion of NMR assignments; Prof. Alexander Katz and Dr. Anh The To (University of California, Berkeley) as well as Dr. Po-Wen Chung (Academia Sinica) for discussion of hydrolysis and adsorption of cellulosic molecules over carbon materials in-depth; Prof. Takashi Kyotani and Dr. Hirotomo Nishihara (Tohoku University) for synthesizing zeolite-templated carbon; Prof. Masa-aki Kakimoto and Dr. Yuta Nabae (Tokyo Institute of Technology) for preparing functionalized polymer; and Mr. Katsuhisa Ishikawa, Mr. Takahiko Hasegawa, and Mr. Tetsuzo Habu (Hokkaido University) for making up quartz-glass instruments.

For reviewing this thesis, Prof. Kazuki Sada, Prof. Koichiro Ishimori, Prof. Keiji Tanino, Prof. Takao Masuda, Prof. Tetsuya Taketsugu, Prof. Kei Murakoshi, Prof. Noboru Kitamura, Prof. Jun-ya Hasegawa, Prof. Masako Kato, Prof. Tamotsu Inabe, Prof. Sadamu Takeda, Prof. Kuniharu Ijiro, Prof. Yukio Hinatsu, Prof. Junji

Nishii (Hokkaido University), and Dr. Kenji Hara (Tokyo University of Technology) have given me constructive comments and suggestions. I gratefully acknowledge Dr. Hara and Dr. Hirokazu Kobayashi in the Fukuoka group for giving me an enormous amount of advice and technical help. I also thank the previous and current members for help and assistance, especially Dr. Hidetoshi Ohta, Dr. Tasuku Komanoya, and Dr. Abhijit Shrotri for significant discussions, and Mr. Shogo Ito and Mr. Kyoichi Kuroki for contributing to this thesis work.

I would like to acknowledge my friends, Mr. Hiroyoshi Fujii, Mr. Ippei Fujisawa, Mr. Naoki Iyo, Mr. Masaru Iwakura, Mr. Tatsuya Matsunaga, Mr. Shunjiro Nagata, Ms. Yukari Sekine, and Mr. Sho Iihama who have motivated each other since 2007.

I have received generous support from the Japan Society for the Promotion of Science (JSPS) Research Fellowship DC2 from April 2014 to March 2015 as well as the JSPS Research Fellowship PD from April 2015 to March 2016. This work was financially supported by a Grant-in-Aid for Scientific Research (KAKENHI, 20226016), a Grant-in-Aid for Young Scientists (B) (KAKENHI, 23760734), a Grant-in-Aid for Young Scientists (A) (KAKENHI, 26709060), and a Grant-in-Aid for JSPS Fellows by JSPS, and the Japan Science and Technology Agency "Advanced Low Carbon Technology Research and Development Program" (JST-ALCA).

Finally, I would like to express my cordial appreciation to my family, Mr. Hiromi Yabushita, Ms. Nanae Yabushita, and Ms. Natsuki Yabushita. They have given me a great amount of support in spirit, and there is no doubt that I have spent unforgettable times and had invaluable experiences in Hokkaido University for eight years owing to their warm encouragement and help.

Berkeley, USA   Mizuho Yabushita
October 2015

Contents

Abbreviations

3A5AF	3-Acetamido-5-acetylfuran
3-HPA	3-Hydroxypropionic acid
5-HMF	5-Hydroxymethylfurfural
AC	Activated carbon
ACC	Aqueous counter collision
ADP	Adenosine diphosphate
AFEX	Ammonia fiber explosion
AH1	Unidentified mono-anhydrohexitol 1
AH2	Unidentified mono-anhydrohexitol 2
APCI	Atmospheric pressure chemical ionization
ATP	Adenosine triphosphate
B3LYP	Becke 3-parameter Lee-Yang-Parr
BET	Brunauer-Emmett-Teller
[BMIM]Cl	1-Butyl-3-methylimidazolium chloride
CBH	Cellobiohydrolase
CNF	Carbon nanofiber
CNT	Carbon nanotube
CP/MAS	Cross polarization/magic angle spinning
CP-SO$_3$H	Sulfonic chloromethyl polystyrene resin
CrI	Crystallinity index
CVD	Chemical vapor deposition
DEPT	Distortionless enhancement by polarization transfer
DF	Degree of freedom
DFT	Density functional theory
DMAc	*N,N*-Dimethylacetamide
DMF	2,5-Dimethylfuran
DMSO	Dimethyl sulfoxide
DOE	Department of Energy
DP	Degree of polymerization
DRIFT	Diffuse reflectance infrared Fourier transform
DTGS	Deuterated triglycine sulfate

EDX	Energy dispersive X-ray spectroscopy
EG	Endoglucanase
FDCA	2,5-Furandicarboxylic acid
GC	Gas chromatography
GlcN	Glucosamine
GlcNAc	N-Acetylglucosamine
GO	Graphene oxide
HAADP	1-O-(2-Hydroxyethyl)-2-acetamido-2-deoxyglucopyranoside
HADP	1-O-(2-Hydroxyethyl)-2-amino-2-deoxyglucopyranoside
HMBC	Heteronuclear multiple bond correlation
HMQC	Heteronuclear multiple quantum coherence
HPA	Heteropoly acid
HPLC	High-performance liquid chromatography
ICIS	International Chemical Information Service
ICP-AES	Inductivity coupled plasma atomic emission spectroscopy
IL	Ionic liquid
IR	Infrared
LC/MS	Liquid chromatography/mass spectroscopy
MC	Mesoporous carbon
MCT	Mercury-cadmium-telluride
MeGlc	1-O-Methylglucose
MeGlcNAc	1-O-Methyl-N-acetylglucosamine
NAD$^+$	Nicotinamide adenine dinucleotide
NADH	Reduced form of nicotinamide adenine dinucleotide
NLDFT	Nonlocal density functional theory
NMR	Nuclear magnetic resonance
NREL	National Renewable Energy Laboratory
OFG	Oxygenated functional group
P(3HB)	Poly(3-hydroxybutyrate)
P(3HB-co-3HV)	Poly(3-hydroxybutyrate-co-3-hydroxyvalerate)
PEIT	Polyethylene isosorbide terephthalate
PET	Polyethylene terephthalate
PTFE	Polytetrafluoroethylene
PVDF	Polyvinylidene difluoride
RI	Refractive index
S/C	Substrate/catalyst
SCRF	Self-consistent reaction field
SEM	Scanning electron microscope
SZ	Sulfated zirconia
TEM	Transmission electron microscope
TEOS	Tetraethyl orthosilicate
TOC	Total organic carbon
TOF	Turnover frequency
TON	Turnover number
TOSS	Total suppression of sidebands

UV	Ultraviolet
XAFS	X-ray absorption fine structure
XRD	X-ray diffraction
YAG	Yttrium–aluminum–garnet
ZTC	Zeolite-templated carbon

Chapter 1
General Introduction

1.1 General Background: Why Non-Food Biomass?

The human civilization has drastically progressed by consuming a massive quantity of fossil fuels as sources of both energy and chemicals since the Industrial Revolution in the eighteenth–nineteenth centuries. Meanwhile, the limitless consumption of fossil fuels has caused worldwide problems such as global warming by greenhouse gases, decrease of pH of ocean, and acid rain. The situation of energy production using nuclear energy has totally changed since March 11, 2011, when a tragic accident happened at a nuclear power plant in Fukushima Prefecture in Japan. Hence, development of new technologies employing safe, environmentally benign, and renewable resources is an enormous challenge to overcome the current issues of energy and chemical production.

Biomass resource, organic compounds produced by animals and plants, is a promising alternative to petroleum owing to its renewability and large quantity, making biorefinery, using biomass as feedstock of fuels and chemicals, an attractive technology to build up sustainable societies. To date, 10^7–10^8 tons of bioethanol has been annually manufactured from corn starch and sugarcane molasses in the United States and Brazil for a decade [1]. However, use of these edible crops as resources competes with food supply, and thus non-food biomass is a more favorable feedstock in the chemical industry. Specifically, cellulose and chitin are attractive resources because of their abundance and potential transformation to various kinds of chemicals [2–10]. Hence, chemical transformation of cellulose and chitin is expected to be a game-changing and beneficial technology for green and sustainable chemistry. In this thesis, the author has aimed at development of new catalytic systems for efficient utilization of both cellulose and chitin as feedstock of platform chemicals.

© Springer Science+Business Media Singapore 2016

M. Yabushita, *A Study on Catalytic Conversion of Non-Food*
Biomass into Chemicals, Springer Theses, DOI 10.1007/978-981-10-0332-5_1

1.2 Cellulose

1.2.1 Plant-Derived Biomass: Lignocellulose

The major components of plant-derived biomass are cellulose (30–50 %), hemicellulose (10–40 %), and lignin (5–30 %) (Fig. 1.1) [11, 12]. These three polymers are intricately intertwined with each other, forming a rigid fiber called lignocellulose, and constitute cell walls in plants. Cellulose is a polymer of glucose units (Fig. 1.2a) linked by β-1,4-glycosidic bonds. Hemicellulose is a copolymer of several kinds of sugar molecules, and a typical building block is β-xylose (Fig. 1.2b). Arabinose (Fig. 1.2c) is sometimes contained in hemicellulose. C6 sugars such as glucose, mannose, fructose, and galactose are also found as minor components in hemicellulose [13, 14]. The structure of hemicellulose depends on plants. Lignin is a 3D-network polymer and its structure depends on plants as well as hemicellulose, making the analysis of lignin structure extremely difficult. It is suggested that three kinds of aromatic compounds, i.e., p-coumaryl alcohol, coniferyl alcohol, and sinapyl alcohol (Fig. 1.3), link with each other by ether bonds to form lignin [15].

Transformation of cellulose, hemicellulose, and lignin to useful compounds is the mainstream of next-generation biorefinery owing to their abundance. Indeed, a variety of reactions to produce various chemicals from lignocellulose, e.g., pyrolysis, hydrolysis, hydrolytic hydrogenation, and hydrodeoxygenation, have been reported [8, 16–18]. However, utilization of lignocellulose has still been a grand challenge due to its recalcitrance. In this section, the author has focused on cellulose, the largest component in lignocellulose, and summarized its characteristics as well as reported catalytic transformation systems.

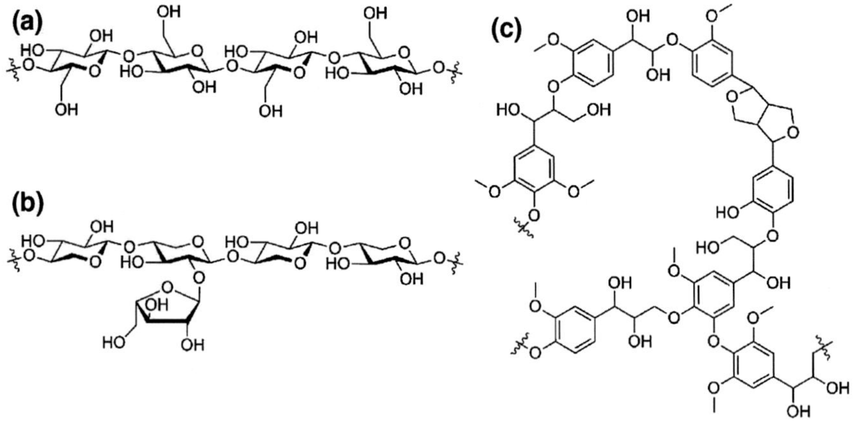

Fig. 1.1 Structures of plant-derived biomass: **a** cellulose; **b** hemicellulose; and **c** lignin

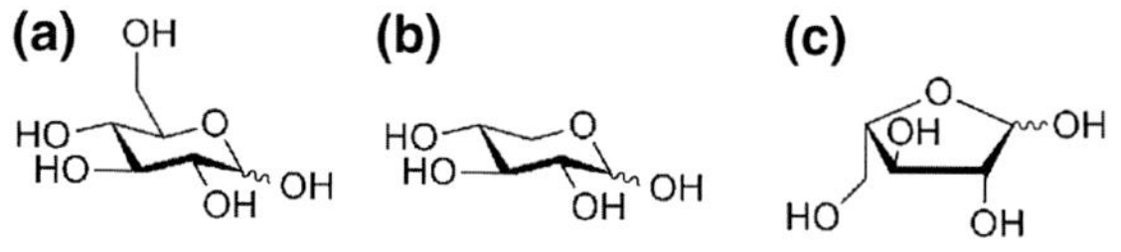

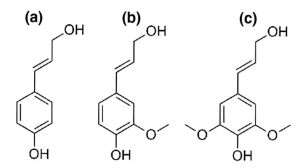

Fig. 1.2 Building blocks of cellulose and hemicellulose: **a** glucose; **b** xylose; and **c** arabinose

(a) OH

(b) OH

(c) OH

Fig. 1.3 Structures of proposed lignin components: **a** *p*-coumaryl alcohol; **b** coniferyl alcohol; and **c** sinapyl alcohol

1.2.2 Crystal Structure of Cellulose

Each glucose unit in cellulose molecule has three hydroxyl groups, which form inter- and intramolecular hydrogen bond network with those of other glucose residues. Consequently, cellulose molecules are uniformly packed; in other words, crystal structure of cellulose is formed. There are seven types of crystal structure, namely I_α, I_β, II, III_I, III_{II}, IV_I, and IV_{II}, which can be identified by powder X-ray diffraction (XRD) and solid-state ^{13}C nuclear magnetic resonance (NMR) spectroscopy [19, 20]. Plants produce cellulose I_β, and bacteria have cellulose I_α. In the crystal structure of cellulose I, cellulose molecules are packed in parallel (so-called parallel chain packing, Fig. 1.4a) [19], and electron diffraction has revealed that I_α and I_β are composed of one-chain triclinic and two-chain monoclinic unit cell, respectively [21]. Other types are artificially obtained by chemical treatments of cellulose I. The mercerization using concentrated NaOH produces type II of crystalline cellulose from type I [22], and cellulose molecules in type II juxtapose with each other in an antiparallel manner (so-called antiparallel chain packing, Fig. 1.4b) [19]. Cellulose III_I and III_{II} are obtained from types I and II, respectively, by a treatment with liquid ammonia [23]. The difference between the types III_I and III_{II} appears to be parallel and antiparallel chain packing, reflecting their respective original forms [19]. Cellulose IV is prepared by treating cellulose III in glycerol at 533 K [23]. Although both XRD and ^{13}C NMR spectroscopy cannot distinguish IV_I and IV_{II} directly, acetylation of hydroxyl groups of cellulose enables to identify them. Cellulose triacetate with parallel chain packing is

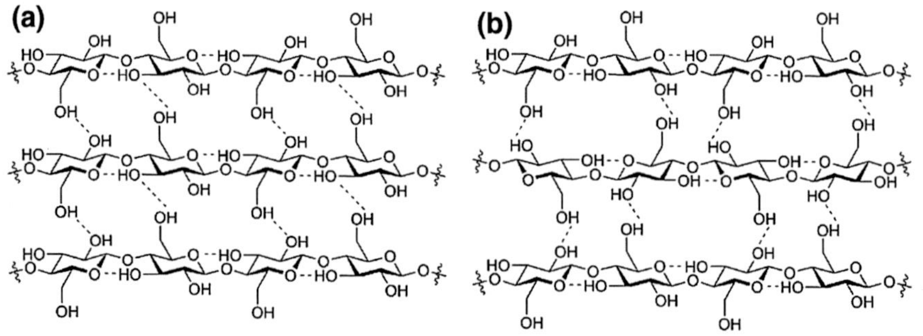

Fig. 1.4 Packing manners of **a** cellulose I (parallel) and **b** cellulose II (antiparallel)

synthesized from cellulose IV_I, and that with antiparallel chain packing is from cellulose IV_{II} [19].

The crystalline structures impart good chemical and physical stability to cellulose. High stability enables to use cellulose in a wide range of applications such as fibers (e.g., rayon) [24], papers, food additives [25], and catalyst supports [26, 27]. In some cases, cellulose nanofiber is favorable for practical application due to its high mechanical strength (equivalent to stainless steel) in addition to five times lower density than steel [28]. The aqueous counter collision (ACC) method is useful for preparation of cellulose nanofiber. This method drastically reduces the length of cellulose particles from 100 μm to 100 nm and the width from 10 μm to 15 nm [29]; meanwhile, the crystal structure of cellulose is maintained.

In stark contrast to the use of cellulose itself, the robust crystalline structure of cellulose is a drawback to its chemical transformation since the access of both reactants and catalysts to reaction sites of cellulose, e.g., glycosidic bonds, has been limited [30]. Crystallinity index (*CrI*) is a useful value to estimate the reactivity of cellulose materials, and cellulose with low *CrI* is desirable for chemical transformation. *CrI* for a mixture of crystalline cellulose I and amorphous cellulose can be determined by XRD (Eq. 1.1, Fig. 1.5) [31], solid-state ^{13}C NMR spectroscopy (Eq. 1.2, Fig. 1.6) [32], and infrared spectroscopy (IR, Eq. 1.3) [33]. XRD and ^{13}C NMR are commonly used and ^{13}C NMR is the most reliable method as the peaks derived from crystalline and amorphous parts are clearly separated [34].

$$CrI_{XRD} = \frac{I_{22.5} - I_{18.5}}{I_{22.5}} \times 100 \tag{1.1}$$

where $I_{22.5}$ and $I_{18.5}$ are intensities observed at 22.5° and 18.5° in XRD pattern, respectively.

$$CrI_{NMR} = \frac{A_{79-86}}{A_{86-92} + A_{79-86}} \times 100 \tag{1.2}$$

Fig. 1.5 XRD patterns of crystalline and amorphous cellulose

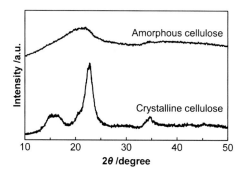

Fig. 1.6 ^{13}C NMR spectra of crystalline and amorphous cellulose. In the figure, C4$_{cr}$ and C6$_{cr}$ peaks are derived from crystal part, and C4$_{am}$ and C6$_{am}$ are from amorphous part

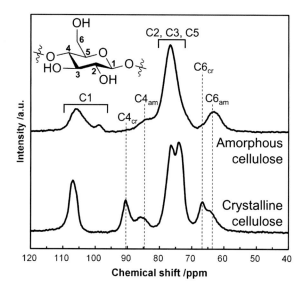

where A_{79-86} and A_{86-92} are peak areas from 79 to 86 ppm and from 86 to 92 ppm in ^{13}C NMR spectrum, respectively.

$$CrI_{IR} = \frac{\alpha_{1372}}{\alpha_{2900}} \times 100 \qquad (1.3)$$

where α_{1372} and α_{2900} are intensities at 1372 and 2900 cm^{-1} in IR spectrum, respectively.

Mechanical and chemical treatments can break the crystalline structure of cellulose by dividing inter- and intramolecular hydrogen bonds, resulting in the decrease of CrI. Mazeau et al. report that each glucose unit in crystalline cellulose I$_\alpha$ and I$_\beta$ form 8 and 6.5 inter- and intramolecular hydrogen bonds, and this number diminishes to on average 5.3 in amorphous cellulose [35]. On a laboratory scale, milling methods such as ball-milling [36] are commonly used due to their ease of handling. For example, ball-milling in a ceramic pot (900 mL) in the presence of ZrO_2 balls (1.0 cm, 1 kg) for 96 h decreased CrI_{XRD} from 80 to 10 %, calculated from Eq. 1.1 and the XRD patterns (Fig. 1.5) [37]. A larger scale of ball-milling [3 L pot with Al_2O_3 balls (1.5 cm, 2 kg)] can reduce the treatment time to 48 h to obtain a similar result. Besides, high-power milling methods such as planetary ball-milling and rod-milling are more time-efficient processes and convert crystalline cellulose into an amorphous one within 1 h [38, 39]. In chemical methods, the treatment of cellulose in H_3PO_4 decreases CrI [40], in which phosphate species possibly form an adduct with cellulosic molecule and weaken their inter- and intramolecular hydrogen bonds [41]. The resulting CrI values depend on the treatment conditions: e.g., CrI changed from 85 to 79 % in 43 % H_3PO_4 at 298 K for 1 h and to 33 % at 323 K for 40 min. During this acid treatment, cellulose is partially hydrolyzed and the degree of polymerization (DP) decreases. Tsao et al. demonstrated the treatment using supercritical CO_2, reducing CrI to ca. 50 % for 1 h [42]. Other physicochemical methods such as ammonia fiber explosion (AFEX) and SO_2-steam explosion are also available [43]. The amorphous cellulose thus prepared is expected to show higher reactivity than the crystalline one [44].

Dissolution of cellulose is another method to improve its reactivity [45], as dissolved cellulose is expected to completely lose the crystalline structure. Cellulose is almost insoluble in water [46] except for at high temperatures with extremely high pressures (e.g., 603 K, 345 MPa) [47]. In contrast to water, lithium chloride/N,N-dimethylacetamide (LiCl/DMAc) dissolves cellulose well, where Cl^- species cleaves inter- and intramolecular hydrogen bonds of cellulose [48, 49]. Schweizer's reagent [$Cu(NH_3)_4(H_2O)_2$](OH)$_2$ also dissolves cellulose since the Cu^{2+} species forms a complex with cellulose and divides hydrogen bonds [50]. These two solvents are well known and are widely used in the field of cellulose research and industry [48, 51]. In 2002, Rogers et al. found new solvents for cellulose, ionic liquids (ILs) such as 1-butyl-3-methylimidazolium chloride ([BMIM]Cl) [52]. This report fascinated many researchers to use a variety of ILs as cellulose solvents due to their chemical and thermal stability as well as tunability of physicochemical properties [53]. Besides, it is known that vapor pressure of ILs is almost zero, which enables to conduct reactions easily even in common vessels; in other words, high pressure reactors are not necessary. The dissolution mechanism of cellulose into ILs is almost the same as that into LiCl/DMAc [53].

1.2.3　Glucose and Its Derivatives

A monomeric unit of cellulose is glucose, which can be produced by hydrolysis. Glucose is one of the energy sources for living things and is used as a sweetener and a medicine against hypoglycemia. In addition to these applications, glucose is a key intermediate to synthesize a variety of chemicals as depicted in Fig. 1.7 [37].

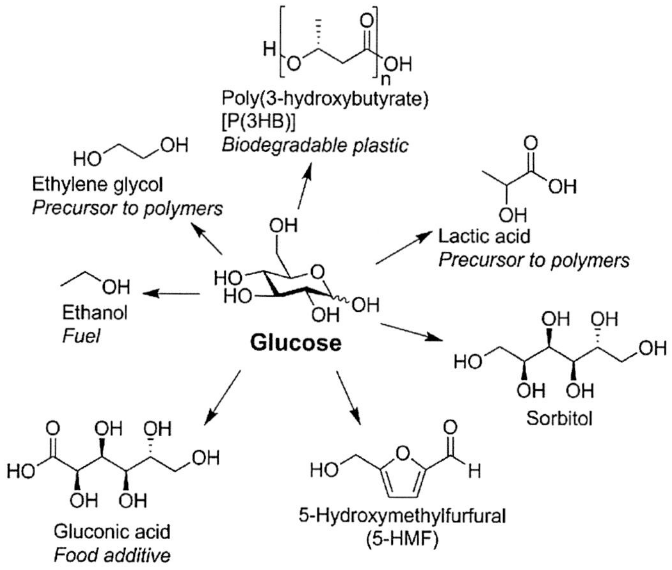

Fig. 1.7 Derivatization of glucose. Further transformations of sorbitol and 5-HMF are depicted in Figs. 1.10 and 1.11, respectively

The present mainstream of chemical transformation of glucose in the world is ethanol production by fermentation via catabolism, including the Embden-Meyerhof-Parnas pathway and successive ethanol fermentation (Fig. 1.8) [54, 55]. During fermentation from pyruvate to ethanol, one equivalent mole of CO_2 is emitted. Ethanol produced from crops is called bioethanol, and the annual production volume has gradually increased in the world [1]. Bioethanol is mainly used as an additive to gasoline, e.g., E15 (including ethanol 15 % and gasoline 85 %) [56], and is now sold in the United States and European countries.

Glucose is a good precursor to polymers owing to its high content of oxygenated functional groups (OFGs). Ethylene glycol, one of the building blocks of polyethylene terephthalate (PET), is produced from glucose by retro-aldol reaction and subsequent hydrogenation. For this transformation, the combination of tungsten (for retro-aldol reaction) and ruthenium (for hydrogenation) species shows high catalytic performance [57]. Taguchi et al. demonstrated the biochemical synthesis of biodegradable plastics, poly(3-hydroxybutyrate) [P(3HB)] and poly (3-hydroxybutylate-*co*-3-hydroxyvalerate) [P(3HB-*co*-3HV)], from glucose by using the recombinant *Escherichia coli* [58, 59].

Glucose is easily hydrogenated to sorbitol by supported metal catalysts [60–63], and this process has been established in the chemical industry. Sorbitol is one of the top 12 building block chemicals produced from biomass resources (Fig. 1.9), suggested by the Department of Energy (DOE) in the United States [64]. This hexitol is

Fig. 1.8 Production scheme
of ethanol from glucose via
Embden-Meyerhof-Parnas
pathway from glucose to
pyruvate and successive
fermentation from pyruvate to
ethanol. Two pyruvate
molecules are produced from
one glucose molecule. *ATP*
adenosine triphosphate; *ADP*
adenosine diphosphate; *NAD⁺*
nicotinamide adenine
dinucleotide; and *NADH*
reduced form of NAD⁺

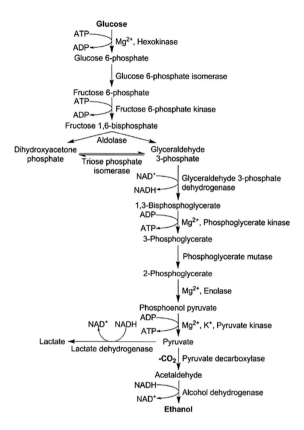

a good precursor to valuable chemicals as well as glucose (Fig. 1.10) [65]. For
example, *n*-hexane and (*E*)-hexatriene are obtained by complete removal of
hydroxyl groups [66, 67]. Sorbitol undergoes dehydration in the presence of acid
catalysts to give two cyclized products, 1,4-anhydrosorbitol (1,4-sorbitan) and
1,4:3,6-dianhydrosorbitol (isosorbide). 1,4-Sorbitan is a raw material for various
environmentally benign surfactants, e.g., fatty acid esters (Span) and polysorbates
(Tween), which are useful as emulsifiers, insecticides, and soft templates for material
synthesis [68]. The emulsifiers are applied to food, cosmetic, and pharmaceutical
industries owing to their low-toxicity to humans [69]. Isosorbide itself and isosorbide dinitrate are medicines for Meniere's disease and angina pectoris, respectively
[70]. Isosorbide-based polymers are outstanding engineering plastics showing good
transparencies and high glass transition temperatures; Mitsubishi Chemical, Teijin,
and Sabic provide isosorbide-based polycarbonates [71–73]. Other polymers such as
polyethylene isosorbide terephthalate (PEIT) are also under consideration as an
alternative to PET [74]. To date, the dehydration of sorbitol has been conducted by
using H_2SO_4 in industrial processes [75]. H_2SO_4 can control the ratio of 1,4-sorbitan
and isosorbide by changing reaction conditions. For example, 58 % yield of
1,4-sorbitan and 18 % yield of isosorbide are formed for 18 min (Table 1.1, entry 1),

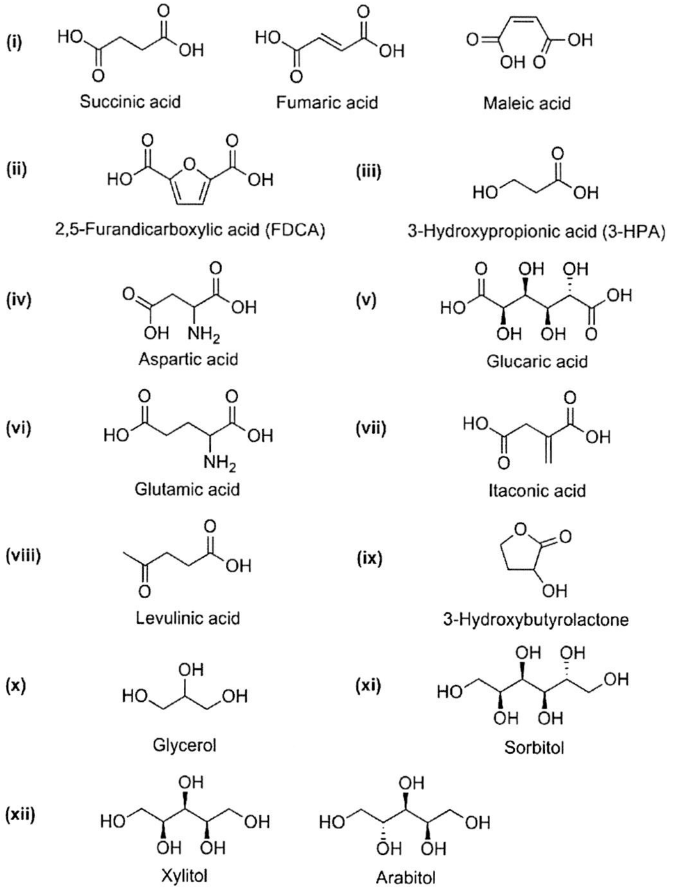

(i) Succinic acid Fumaric acid Maleic acid

(ii) 2,5-Furandicarboxylic acid (FDCA) (iii) 3-Hydroxypropionic acid (3-HPA)

(iv) Aspartic acid (v) Glucaric acid

(vi) Glutamic acid (vii) Itaconic acid

(viii) Levulinic acid (ix) 3-Hydroxybutyrolactone

(x) Glycerol (xi) Sorbitol

(xii) Xylitol Arabitol

Fig. 1.9 Top 12 building block chemicals produced from biomass resources, suggested by DOE in the United States [64]

whereas isosorbide (66 %) becomes the major product for 48 min (entry 2). It is noteworthy that the details of the catalysis for sorbitol dehydration by H_2SO_4 have not yet been revealed. Recently, a variety of heterogeneous catalysts have been developed to produce isosorbide from sorbitol on a laboratory scale (Table 1.1, entries 3–13), e.g., immobilized $H_3PW_{12}O_{40}$ [76], metal phosphate [77], $CuSO_4$ [78], Ru modified Raney Cu [79], sulfonated polystyrene resin (Amberlyst 35) [80], sulfonated oxides [81, 82], and zeolites [83–86]. Heterogeneous catalysts are beneficial to industrial processes because they are easily separated from reaction products and recycled. Although the selective production of isosorbide by heterogeneous catalysts has been reported as shown above, that of 1,4-sorbitan still remains a grand challenge since 1,4-sorbitan readily undergoes subsequent dehydration to form isosorbide in the presence of acids. Takagaki et al. reported that

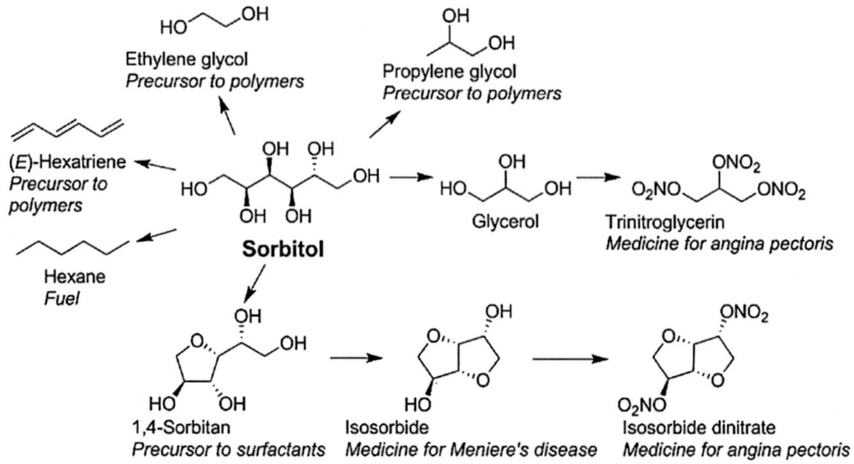

Fig. 1.10 Derivatization of sorbitol

Table 1.1 Dehydration of sorbitol to 1,4-sorbitan and isosorbide by acid catalysts

Entry	Catalyst	T/K	Time/h	Conv./%	Yield/%		References
					1,4-Srb[a]	Iso[b]	
1	H_2SO_4	453	0.3	98	58	18	[75]
2	H_2SO_4	453	0.8	100	0.1	66	[75]
3	$H_3PW_{12}O_{40}/SiO_2$	523	0.4	96	16	56	[76]
4	SnPO	573	2	72	24	47	[77]
5	$CuSO_4$	473	4	100	n.d.[c]	67	[78]
6	Ru–Raney Cu	533	4	n.d.[c]	7.5	51	[79]
7	Amberlyst 35	408	2	n.d.[c]	19	73	[80]
8	Sulfated ZrO_2	483	2	100	18	65	[81]
9	Sulfated TiO_2	503	4	100	13	75	[82]
10	H–MFI[d]	533	1	100	n.d.[c]	50	[83]
11	H–MOR[d]	523	5.8	85	n.d.[c]	40	[84]
12	H–BEA[e]	400	2	>99	4.6	76	[85]
13	H–BEA[e]	453	60	n.d.[c]	n.d.[c]	81	[86]
14	$HNbMoO_6$	433	60	94	42	21	[87]

[a]1,4-Sorbitan
[b]Isosorbide
[c]No data
[d]Si/Al ratio was not shown
[e]Si/Al ratio = 75

layered niobium molybdate $HNbMoO_6$ provided ca. 40 % yield of 1,4-sorbitan from sorbitol (entry 14) [87]. They have proposed that the key to high-yielding production of 1,4-sorbitan is molecular recognition that can distinguish two substrates,

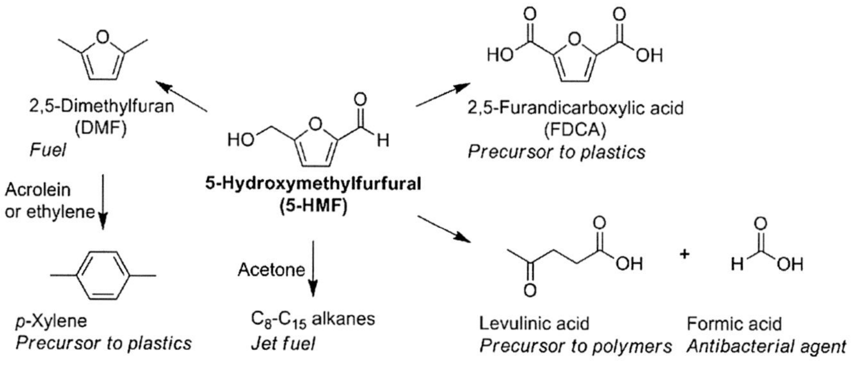

Fig. 1.11 Derivatization of 5-HMF

i.e., sorbitol and 1,4-sorbitan. In other words, selective interaction between catalysts and sorbitol would enable to produce 1,4-sorbitan in good yield and selectivity.

5-Hydroxymethylfurfural (5-HMF) has attracted much attention as 5-HMF derivatives are applicable in many fields (Fig. 1.11). The production of 5-HMF from glucose is as yet under consideration in laboratories, but high-yielding synthesis of 5-HMF has been achieved. ILs and biphasic solvent systems (water and organic solvents) are typical reaction media for this transformation since the produced 5-HMF is effectively extracted to organic solvents and its further degradation in water could be inhibited [88]. Zhang et al. demonstrated the first direct synthesis of 5-HMF from glucose using $CrCl_2$ in ILs [89]. $CrCl_2$ associates with glucose and then accelerates the reaction. Later, mechanistic studies have revealed that the transformation of glucose to 5-HMF requires both Lewis and Brønsted acids, as Lewis acids are necessary for epimerization of glucose to fructose [90] and Brønsted ones are for successive dehydration of fructose to 5-HMF [91]. This insight leads to the development of bifunctional heterogeneous catalysts such as Sn-modified BEA zeolite [92], sulfonated oxides [93], and BEA zeolite [94]. Although these catalytic systems require biphasic systems, Hara et al. synthesized 5-HMF from glucose in 81 % yield in water by using phosphate/TiO_2 as a catalyst [95]. Surprisingly, this phosphate/TiO_2 catalyst works even in water, as typical Lewis acids are quenched in the presence of water molecules. 5-HMF is derivatized to a variety of chemicals as shown in Fig. 1.11. The hydrodeoxygenated product of 5-HMF is 2,5-dimethylfuran (DMF) [2], high-octane number fuel with high energy density (30,000 kJ L^{-1}). DMF is further transformed into p-xylene, a precursor to PET, through Diels-Alder reaction with acrolein or ethylene [96, 97]. The oxidation of 5-HMF over metal catalysts provides 2,5-furandicarboxylic acid (FDCA) [98, 99], which could be a polymer-substrate alternative to terephthalic acid. 5-HMF also undergoes hydrolysis to form levulinic acid and formic acid [100]. Levulinic acid is a good building block of polymers, lubricants, and medicines [101]. Furthermore, Dumesic et al. demonstrated the production of C_8–C_{15} alkanes for jet

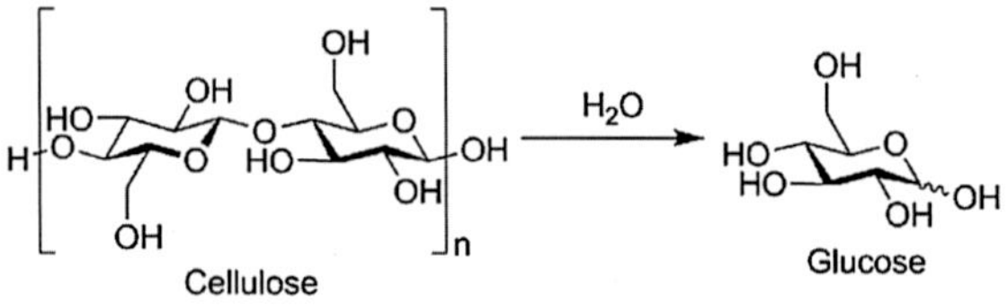

Fig. 1.12 Hydrolysis of cellulose to glucose

fuels by hydrogenation and hydrodeoxygenation following condensation with acetone [102].

1.2.4 Hydrolysis of Cellulose in Homogeneous Systems

Production of glucose from cellulose is one of the target reactions in biorefinery due to the wide applications of glucose as mentioned in Sect. 1.2.3. Hydrolysis of cellulose to glucose (Fig. 1.12) is a grand challenge due to recalcitrance of cellulose, and researchers have made efforts to depolymerize cellulose using various kinds of catalytic systems. In this section, the author summarizes the reported homogeneous catalytic systems to hydrolyze cellulose.

The conventional catalysts for cellulose hydrolysis to glucose are H_2SO_4 and cellulase enzymes [103, 104]. Industrial processes employing an H_2SO_4 aqueous solution are known as the Scholler process using diluted H_2SO_4 (0.5 %) developed in Germany [105] and Hokkaido process employing concentrated H_2SO_4 (30–70 %) operated in Japan [106], and the processes produced glucose in up to 90 % yield. However, these processes are unprofitable due to equipment corrosion by H_2SO_4, complicated separation system of glucose from reaction cocktail, and acidic waste disposal; to date, they have been suspended.

Recently, mechanocatalytic depolymerization of cellulose using acid catalysts, e.g., kaolinite ($Al_2Si_2O_7 \cdot 2H_2O$) and H_2SO_4, without solvents or heat has been reported. In a work by Blair et al., cellulose and kaolinite were planetary ball-milled together at 350 rpm for 3 h, and 84 % of cellulose was depolymerized into water-soluble products [107]. Rinaldi et al. conducted planetary ball-milling of H_2SO_4-impregnated cellulose at 350 rpm for 2 h, and cellulose was almost completely hydrolyzed into water-soluble oligosaccharides [108]. In these systems, the glycosidic bonds activated by acids receive mechanical force to be activated and cleaved. Beltramini et al. characterized the produced oligomers in detail using 2D NMR and found the formation of branched polymers linked by α-1,6-glycosidic bonds, which improved the solubility of the oligomers [109]. This insight indicates that both hydrolysis and condensation of glucans occur during the milling process. No researchers have so far explained the origin of water molecules that react with the substrate to hydrolyze glycosidic bonds and form hydroxyl groups. The author speculates that cellulose undergoes hydrolysis by physisorbed water on the substrate or water in atmosphere. Cellulose generally contains 5–10 wt% of physisorbed water. For example, 1 g of the substrate with 10 wt% of water contains 0.9 g of cellulose (equivalent to 5.6 mmol of glucose unit) and 0.1 g of water (5.6 mmol); hence, the amount of physisorbed water is sufficient to hydrolyze cellulose to oligomers and glucose. Owing to high solubility and reactivity of the produced oligomers, further conversion easily proceeds. Rinaldi et al. reported that 91 % yield of glucose was formed from the water-soluble oligosaccharides in a batch-type reactor [108]. They also conducted hydrolytic hydrogenation of the oligomers to sorbitol under gaseous H_2 and achieved 86 % yield (reaction scheme is shown in Sect. 1.2.6) [110]. Beltramini et al. demonstrated the hydrolytic transfer-hydrogenation of the oligomers to produce sorbitol (reaction scheme is shown in Sect. 1.2.6) by using Ru/C catalyst and 2-propanol as a hydrogen source in a fixed-bed flow reactor due to difficult control of flow of gaseous H_2 and liquid water at the same time on a laboratory scale [111]. The remaining concern of these systems is reusability of the homogeneous catalyst H_2SO_4.

Heteropoly acids (HPAs) have been studied as homogeneous catalysts working in water. HPAs are advantageous over typical mineral acids such as H_2SO_4 and HCl due to their separability by extraction using organic solvents such as diethyl ether. Another advantage of HPAs is high acidity. For example, the Hammet acidity function (H_0) of tungstoboric acid $H_5BW_{12}O_{40}$ (0.7 M at 298 K, vide infra) is −2.1, and this value is lower than those of H_2SO_4 and HCl (0.7 M) (H_0 = ca. 0) [112]. Hence, HPAs are expected to be active and reusable catalysts for hydrolysis instead of mineral acids. Shimizu et al. demonstrated the hydrolysis of ball-milled cellulose by a variety of HPAs [113]. The total yields of reducing sugars from cellulose were 18 and 17 % when using $H_3PW_{12}O_{40}$ and $H_4SiW_{12}O_{40}$ at 423 K for 2 h, respectively, and these values were higher than those given by typical mineral acids, e.g., 12 % for HClO and 10 % for H_2SO_4. An HPA-based salt $Sn_{0.75}PW_{12}O_{40}$ afforded higher glucose yield (23 %) under the same reaction conditions, and the yield was maximized at ca. 40 % at 423 K for 15 h. They further investigated the effects of counter cations of $PW_{12}O_{40}^{3-}$ on cellobiose hydrolysis. A volcano-type correlation was found between turnover frequencies (TOFs) for glucose formation and Lewis acidities, and $Sn_{0.75}PW_{12}O_{40}$, $RuPW_{12}O_{40}$, and $ScPW_{12}O_{40}$ provided high TOFs in their tests. Thus, both Brønsted and Lewis acidities of HPAs play significant roles in hydrolysis of glycosidic bonds. Mizuno et al. found that $H_5BW_{12}O_{40}$ showed remarkably higher catalytic performance even for hydrolysis of crystalline cellulose (77 % yield of glucose) than $H_3PW_{12}O_{40}$ (8 %) and $H_4SiW_{12}O_{40}$ (37 %) [112]. They mentioned that the acidity and function of decreasing CrI of cellulose were important for catalytic activity of HPAs. In fact, the order of acidity was $H_3PW_{12}O_{40}$ < $H_4SiW_{12}O_{40}$ < $H_5BW_{12}O_{40}$, which was the same as the order of catalytic activity. Their further study showed that HPAs composed of highly negatively charged anions were preferable and such anions clearly decreased CrI. In other words, highly negatively charged anions possibly work as good hydrogen-bond acceptors and weaken inter- and intramolecular hydrogen bonds of cellulose. Besides, protons of HPAs also weakened hydrogen bonds of cellulose, and a higher concentration of proton was suitable for this purpose. Thus,

Fig. 1.13 Catalytic domain
of CBH, in which a cellulosic
molecule penetrates into a
tunnel. Reprinted from Ref.
[117], Copyright 1998, with
the permission from Elsevier

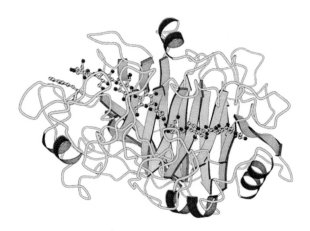

$H_5BW_{12}O_{40}$ worked as a highly active catalyst for the hydrolysis of cellulose. $H_5BW_{12}O_{40}$ can be recovered by extraction and recycled for 10 times.

In enzymatic hydrolysis of cellulose to glucose, three types of enzymes, i.e., endoglucanase (EG), exoglucanase (so-called cellobiohydrolase, CBH), and β-glucosidase, are simultaneously employed [114]. EG randomly cleaves β-1,4-glycosidic bonds of cellulose in amorphous parts to increase the number of reducing/non-reducing ends. Subsequently, CBH depolymerizes cellulose from the reducing/non-reducing ends in both crystalline and amorphous parts to produce only cellobiose. β-Glucosidase works for the hydrolysis of soluble oligomers produced by EG and CBH to glucose. The functions of EG and CBH for the hydrolysis of cellulose are proposed as follows. First, enzymes adsorb onto the surface of cellulose through hydrophobic interactions derived from the binding domain composed of three exposed tyrosine residues [115]. Second, cellulose chains are introduced into a pocket (in the case of EG) or a tunnel (CBH, Fig. 1.13), where there is an active site [116, 117]. Then a glycosidic bond in cellulose is cleaved by a pair of carboxylic acid and carboxylate, in which two types of mechanism, inverting and retaining ones, are presumed (Fig. 1.14) [118]. Nowadays, cellulase enzymes can afford glucose in 75 % yield from cellulose under moderate conditions (e.g., 343 K, 1.5 days) [103, 104], and some enzymatic systems have been extended to a commercial scale. Although adsorption process separates EG and CBH from a product solution after the reaction, β-glucosidase does not adsorb on the adsorbent and thus cannot be recovered [119, 120]. Immobilization of cellulases has been considered for easy separation and recycle, and various reports have demonstrated recycle tests of the catalysts [121–123]. However, the immobilized cellulases suffer from their less catalytic performances than that of *free* cellulases because of low accessibility of the substrates to the active sites.

Reaction systems using only supercritical water as both solvent and reactant are tested in laboratories and have produced glucose from cellulose in 24 % yield [124, 125]. However, a special reactor resisting harsh conditions is required. Another drawback is further degradation of produced glucose at such a high

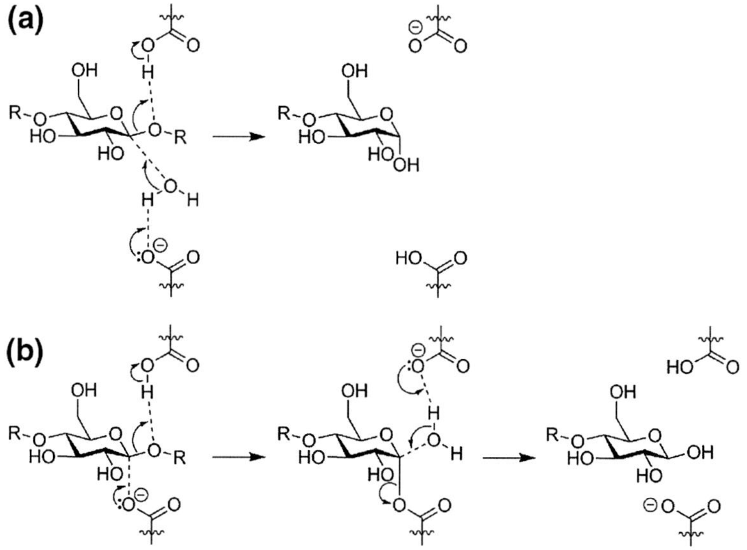

Fig. 1.14 Proposed reaction mechanisms for glycosidic bond cleavage by cellulase enzymes: **a** inverting mechanism and **b** retaining mechanism

temperature (673 K). Gaseous water is also used for the hydrolysis of crystalline cellulose [126]. Water vapor goes into interstices of cellulose molecules, weakens inter- and intramolecular hydrogen bonds, and hydrolyzes glycosidic bonds. Although this reaction system gives glucose in ca. 20 % yield from crystalline cellulose, at issue is low selectivity of glucose due to side reactions.

1.2.5 Hydrolysis of Cellulose by Heterogeneous Catalysts

Heterogeneous catalysis for cellulose hydrolysis has been investigated as heterogeneous catalysts are advantageous over homogeneous ones thanks to their ease of separation and reusability.

As described in Sect. 1.2.2, cellulose dissolved in ILs is a homogeneous substrate and is expected to undergo hydrolysis more easily than cellulose in water. Schüth et al. employed a sulfonated polystyrene resin (Amberlyst 15DRY), which is commonly used as a heterogeneous and strong acid catalyst, in [BMIM]Cl (Table 1.2, entry 15) [127]. Later, it was found that H^+ species of sulfonic group on the resin was exchanged with $[BMIM]^+$ and the released H^+ in the solution catalyzed the hydrolysis [128]. The replaced $[BMIM]^+$ species remained on the resin even after the reaction, causing deactivation of Amberlyst 15DRY. When recycling Amberlyst 15DRY, renegeration by ion exchange with H_2SO_4 is necessary. ILs-immobilized char with acidic groups was developed to easily separate the

Table 1.2 Hydrolysis of cellulose to glucose by heterogeneous catalysts

Entry	Catalyst	Pretreatment of cellulose	S/C^a	T/K	Time/h	Conv./%	Yield/% Glc[b]	Olg[c]	References
15[d]	Amberlyst 15DRY	Microcrystalline	5.0	373	5	29	0.9	0.6[j]	[127]
16[e]	C-SO$_3$H-IL	Microcrystalline	2.0	363	2	n.d.[h]	33[i]	n.d.[h]	[129]
17	C-SO$_3$H	Microcrystalline	0.083	373	3	n.d.[h]	4	64	[131]
18	AC-SO$_3$H	Ball-milled	0.9	423	24	n.d.[h]	41	n.d.[h]	[138]
19	SiO$_2$-C-SO$_3$H	Ball-milled	1.0	423	24	61	50	2	[140]
20	CMK-3-SO$_3$H	Ball-milled	0.9	423	24	94	75	n.d.[h]	[141]
21	Fe$_3$O$_4$-SBA-SO$_3$H	[BMIM]Cl	0.7	423	3	n.d.[h]	50	n.d.[h]	[144]
22	Fe$_3$O$_4$-SBA-SO$_3$H	Microcrystalline	1.0	423	3	n.d.[h]	26	n.d.[h]	[144]
23	CoFe$_2$O$_4$/SiO$_2$-SO$_3$H	[BMIM]Cl	1.0	423	3	n.d.[h]	7.0	30[j]	[146]
24	CP-SO$_3$H	Microcrystalline	0.2	393	10	n.d.[h]	93	n.d.[h]	[147]
25	2.0 wt% Ru/CMK-3	Ball-milled	6.5	503	0.83[f]	56	24	16	[150]
26	CMK-3	Ball-milled	6.5	503	0.83[f]	54	16	5	[150]
27	MC	Ball-milled	3.3	518	0.75[g]	71	41	0.9[j]	[141]
28	GO	Not mentioned	0.9	423	24	59	50	4[j]	[153]

[a]Ratio of substrate and catalyst based on weight
[b]Glucose
[c]Oligosaccharides
[d]The reaction was conducted in [BMIM]Cl
[e]Microwave (350 W) was used in addition to heat
[f]Total time including heating and cooling. The keeping time at 503 K was less than 1 min
[g]The reaction mixture was heated from room temperature to 518 K in 45 min and rapidly cooled down to room temperature
[h]No data
[i]Total yield of reducing sugars
[j]Cellobiose

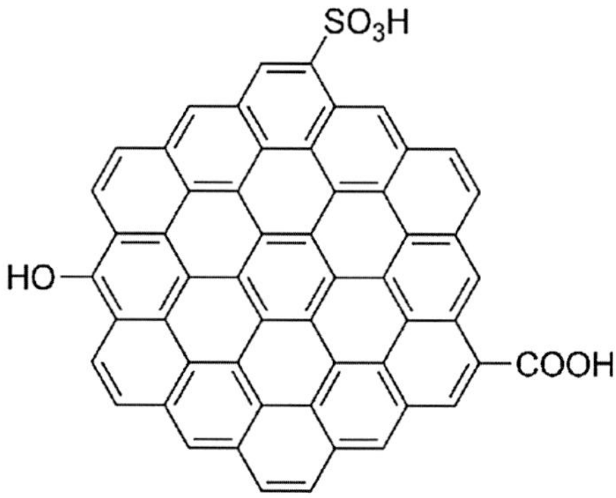

Fig. 1.15 Proposed structure
of sulfonated carbon [131]

catalyst from products [129]. This catalyst worked for the reaction even in water and afforded 33 % yield of glucose (entry 16). The immobilized ILs likely interact with cellulose to weaken hydrogen bond network and enhance reactivity of cellulose. Since ILs are expensive and toxic, a decisive advantage is necessary to use ILs as a solvent instead of water.

For the viewpoint of green chemistry, heterogeneous catalysts working under aqueous conditions are required. H_2SO_4 is a common catalyst for hydrolysis of cellulose (vide supra), and thus the immobilization of sulfonic groups onto solid surfaces have been extensively studied in the past decade. Hara et al. synthesized a sulfonated carbon (Fig. 1.15) by carbonization of cellulose at 723 K under N_2, followed by sulfonation using oleum at 353 K [130]. The resulting sulfonated carbon hydrolyzed microcrystalline cellulose to glucose (4 % yield) and water-soluble oligosaccharides (64 %) in water at 373 K in a Pyrex test tube (entry 17) [131–133]. Although it is known that sulfonic groups on large graphene sheets readily undergo decomposition in hot-compressed water [134], those on this material are resistant to such harsh conditions due to small size of graphene sheets (ca. 1 nm). In addition, carboxylic acids on the edge of graphene sheets stabilize the sulfonic groups by an electron withdrawing effect. As a result, the sulfonated carbon maintained its catalytic activity in reuse tests more than 25 times. For the catalytic mechanism, they proposed that COOH groups as well as neutral or weakly acidic OH groups such as phenolic groups on carbon lead to adsorption of the substrate onto the catalyst surface by forming hydrogen bonds, followed by hydrolysis of glycosidic bonds by sulfonic groups [135, 136]. Hence, OFGs would help the catalytic reaction occurring on the catalyst surface. Later, they also submitted a patent for a specific reactor equipped with a strong agitation wing grinding cellulose–catalyst slurry [137]. The author therefore speculates that mechanical force may degrade the crystalline structure of cellulose during their catalytic reactions, giving a good performance in the hydrolysis of microcrystalline cellulose. Meanwhile, Onda et al. prepared sulfonated carbon (denoted as $AC–SO_3H$) by boiling commercial activated carbon (AC) in concentrated H_2SO_4, and $AC–SO_3H$ produced glucose in 41 % yield at 423 K for 24 h (entry 18) [138, 139]. After exposing AC–SO$_3$H to reaction conditions without cellulose, they performed hydrolysis of cellulose again using the filtrate as a solvent without adding any solid catalysts; as a result, glucose yield was as low as that of a blank reaction without catalyst. In the author's test, a trace amount of sulfate ions (35 µM, corresponding to 1.2 % of sulfonic groups in the catalyst) was detected in the filtrate [37]. Furthermore, AC–SO$_3$H gave similar glucose yield in a reuse test. These results clearly showed that hydrolysis was accelerated by solid AC–SO$_3$H and not by soluble species derived from the catalyst. Note that they treated the material in hot-compressed water at 473 K during its preparation in order to remove weakly bonded sulfonic groups, and the remaining sulfonic groups on AC were stable and reusable at 423 K. This treatment should be crucial for good durability of AC–SO$_3$H.

These pioneering works have motivated other researchers to develop sulfonic acid catalysts to improve activity and separability. A silica/carbon nanocomposite treated by H$_2$SO$_4$ yielded glucose in 50 % (entry 19) [140]. Zhang et al. synthesized sulfonated mesoporous carbon CMK-3, which produced 75 % yield of glucose (entry 20) [141]; this value is one of the highest glucose yield ever reported. They argued that the mesoporous structure of the catalyst led to good mass transfer, resulting in the high catalytic performance. Indeed, Katz et al. revealed that long-chain cello-oligosaccharides rapidly diffused and adsorbed into mesopores of carbon material [142, 143]. Magnetic catalysts with sulfonic groups, e.g., a sulfonated composite of mesoporous silica SBA-15 and Fe$_3$O$_4$ (entries 21 and 22) [144, 145] and sulfonated CoFe$_2$O$_4$-embedded silica (entry 23) [146], were also useful as they were easily separated from a product cocktail by a magnet after the reaction. Pan et al. synthesized a sulfonic chloromethyl polystyrene resin (CP–SO$_3$H, Fig. 1.16) [147]. In their concept, Cl groups are expected to interact with OH groups of cellulosic molecules to adsorb the substrate onto the catalyst surface, followed by the hydrolysis of glycosidic bonds by sulfonic groups. They remarked that CP–SO$_3$H converted microcrystalline cellulose to glucose in 93 % yield at 393 K for 10 h (entry 24). However, the stability of CP–SO$_3$H needs to be reconfirmed. The base polymer chloromethyl polystyrene, known as Merrifield's

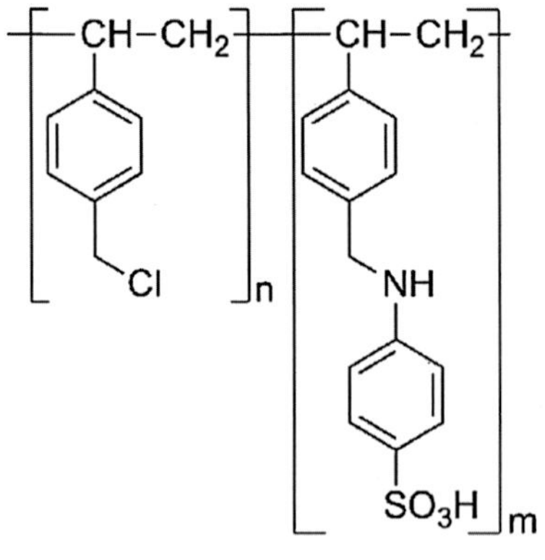

Fig. 1.16 Structure of CP–SO$_3$H. Cl and SO$_3$H groups are proposed to be binding and active sites, respectively [147]

resin, easily undergoes S_N2 reaction at a benzyl position as widely used for connecting peptides and proteins [148]. Indeed, in their synthetic scheme of CP–SO_3H, sulfanilic acid nucleophilically attacks a carbon atom at the benzyl position of Merrifield's resin to form a C–N covalent bond. Besides, leaching of both Cl^- and SO_4^{2-} species was detected by ion chromatography after hydrolysis of cellulose using CP–SO_3H in the author's test, and this was possibly due to S_N2 reaction at the benzyl position of CP–SO_3H by water molecules [37]. It is important to take care of the possibility of leaching of SO_3H and Cl groups in sulfonated or chlorinated materials in water at high temperatures. Furthermore, the interaction between the substrate and catalyst is not clear. Back to their original concept, Cl^- species has good interaction with cellulose [48, 49]; contrastingly, covalently bonded Cl is hardly expected to have a similar ability; the author and co-workers did not observe the interaction between covalently bonded Cl with OH groups of cellobiose [149]. Although these sulfonic acid catalysts afford high glucose yields, substrate/catalyst [S/C (wt/wt)] ratios of the reaction were typically lower than 1; higher S/C ratios are favorable to improve efficiency.

Supported metal catalysts have also been used for the hydrolysis of cellulose. Fukuoka et al. found that 2 wt% Ru/CMK-3 hydrolyzed cellulose to glucose in 24 % yield in water (entry 25) [150, 151], and this catalyst was reusable up to five times without loss of activity or Ru leaching, confirmed by inductivity coupled plasma atomic emission spectroscopy (ICP-AES). The yield of glucose was raised up to 31 % by increasing the Ru loading to 10 wt%. Interestingly, this catalyst promotes the reaction without addition of any acids or any post-synthetic modifications of the CMK-3 support, indicating that the impregnated Ru nanoparticles possibly work as a catalyst for the cleavage of glycosidic bonds. The Ru species on CMK-3 was highly dispersed $RuO_2 \cdot 2H_2O$ characterized by X-ray absorption fine structure (XAFS) spectroscopy and XRD, and the high valent Ru species probably play an important role in the reaction. They proposed that an empty site on Ru atom worked as a Lewis acid and/or a water molecule coordinating to Ru did as a Brønsted acid (Fig. 1.17).

It is known that acids with a pK_a value larger than –3 are usually ineffective for the hydrolysis of cellulose [152]. Surprisingly, mesoporous carbon (MC) and

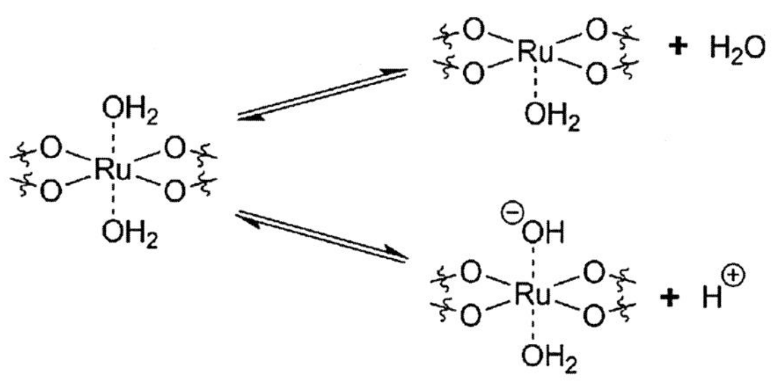

Fig. 1.17 Speculated active sites on Ru/CMK-3 catalyst [151]: a function as Lewis acid (upper route) and Brønsted acid (lower one)

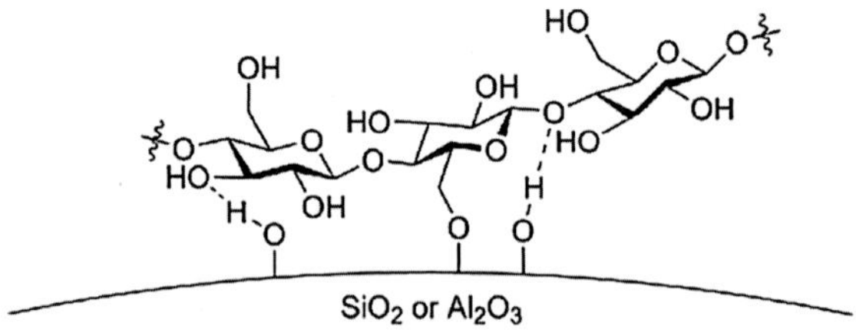

Fig. 1.18 Schematic of activation mechanism of glycosidic bonds in grafted cellulosic molecules [158]

graphene oxide (GO) catalyzed the hydrolysis of β-1,4-glycosidic bonds even in the absence of strong acidic groups or metals (entries 26–28) [141, 150, 151, 153–155]. The weakly acidic groups such as carboxylic acids and phenolic groups are proposed as active sites; however, the detailed structures and their catalysis remain unclear. Katz et al. also reported that even OH groups on SiO_2 and Al_2O_3 with very weak acidity (pK_a = ca. 7) were able to hydrolyze cellulosic molecules [156–158]. This reaction occurred only when inducing stressful conformations by grafting cellulosic molecules onto SiO_2 (and Al_2O_3) surface via the formation of ether bonds between OH groups of the substrate and those on SiO_2 (and Al_2O_3). This enthalpically favored adsorption kept other neighboring OH groups on SiO_2 (and Al_2O_3) close to the immobilized polymer. In this case, the OH groups had enhanced chances to activate glycosidic bonds as depicted in Fig. 1.18. Note that neither SiO_2 nor Al_2O_3 could hydrolyze glycosidic bonds without immobilization of the substrate. These results clearly show the importance of juxtaposing a substrate with surface active sites.

1.2.6 Hydrolytic Hydrogenation of Cellulose to Sorbitol

Direct conversion of cellulose into platform chemicals other than glucose has also been investigated; notably, hydrolytic hydrogenation of cellulose to sorbitol (Fig. 1.19) is an active area in biorefinery. The author therefore introduces the representative works in this section (Table 1.3).

In the 1950s, Balandin et al. demonstrated the hydrolytic hydrogenation of cellulose to sorbitol by Ru/C in an H_2SO_4 aqueous solution under 7 MPa of H_2 (entry 29) [159, 160]. Soluble H_2SO_4 hydrolyzes insoluble cellulose to glucose, which undergoes hydrogenation to sorbitol over solid Ru/C. In detail, formation of soluble oligosaccharides and their reduced alcohols are involved in the reaction [161, 162]; however, they are omitted for better handling and easier understanding the reaction. In 1989, a bifunctional catalyst, Pt over FAU zeolite, was developed for the hydrolytic hydrogenation of starch [163]. Starch composed of α-glucose units is soluble in water, making its chemical transformation significantly easier than cellulose conversion. Solid FAU zeolite hydrolyzes soluble starch to glucose

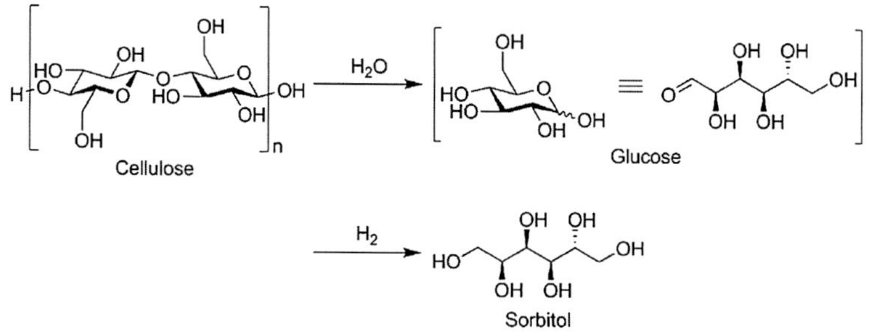

Fig. 1.19 Hydrolytic hydrogenation of cellulose to sorbitol via glucose

and solid Pt nanoparticles hydrogenate glucose to sorbitol under H_2 [164]. Both steps are solid–liquid interfacial reactions, and thus the conversion of *solid* cellulose by *solid* catalysts has remained a concern for a long time.

In 2006, Fukuoka et al. reported the first hydrolytic hydrogenation of cellulose using only solid Pt/γ-Al$_2$O$_3$ catalyst, which produced sorbitol and mannitol, an isomer of sorbitol, in 25 and 6 % yields, respectively (entry 30) [165]. Intriguingly, the hydrolysis step is also accelerated by the solid catalyst regardless of low acidity of γ-Al$_2$O$_3$ (pK_a = 3.3–6.8 [166]), which cannot hydrolyze *free* cellulose dispersed in water [158]. Although it was proposed that the hydrolysis was accelerated by proton produced via heterolysis of H_2 over Pt nanoparticles [165, 167, 168], this mechanism would be excluded since such species were not observed in pyridine-adsorbed IR spectroscopy [169]. As yet, the mechanism of the reaction over supported Pt catalysts has not been revealed. The residual Cl on the catalysts derived from catalyst precursors such as H$_2$PtCl$_6$ probably leads to the formation of HCl or AlCl$_3$. HCl and Al^{3+} species enhanced the rates of cellulose hydrolysis and side reactions [170, 171]. This problem is avoidable by using Cl-free precursors such as Pt(NH$_3$)$_2$(NO$_2$)$_2$ and carbon supports. Liu et al. argued that protons from hot compressed water promoted the hydrolysis step [172, 173] due to the change in pK_w of water $[-\log(a_{H^+} \cdot a_{OH^-})]$ to 11.2 at 448 K [174]. However, its corresponding pH (5.6) is too high to catalyze hydrolysis of cellulose. The direct attack of water molecule to glycosidic bonds may be more significant [175]. Further works are necessary to scrutinize real active species.

Later, γ-Al$_2$O$_3$ support was found to be transformed into boehmite AlO(OH) in hot compressed water, causing the drop of activity in repeated tests [170]. In addition to poor durability, yield of sorbitol needs to be improved. Thus, the development of more active and durable catalysts has been a target. Carbon materials such as AC and carbon nanotubes (CNTs) are known as heat- and water-tolerant supports. For this reason, carbon supported metal catalysts have been extensively studied. Liu et al. conducted the hydrolytic hydrogenation of cellulose by a durable Ru/C catalyst, which produced hexitols in 39 % yield with 86 % conversion of cellulose (entry 31) [173]. Neither leaching nor sintering of Ru nanoparticles, both of which are typical deactivation reasons for supported metal

Table 1.3 Hydrolytic hydrogenation of cellulose to sugar alcohols

Entry	Catalyst	Pretreatment of cellulose	T/K	Time/h	$P(H_2)$/MPa	Conv./%	Yield/%			References
							Sor[a]	Man[b]	Total[c]	
29	H_2SO_4, Ru/C	Sulfite[h]	433	2	7	n.d.[k]	n.d.[k]	n.d.[k]	82[l]	[159]
30	2.5 wt% Pt/γ-Al_2O_3	Microcrystalline	463	24	5.0	n.d.[k]	25	6	31	[165]
31	4.0 wt% Ru/C	Microcrystalline	518	0.5	6.0	86	30	10	39	[173]
32	1.0 wt% Ru/CNT	H_3PO_4	458	24	5.0	n.d.[k]	69	4	73	[40]
33	1.0 wt% Ru/CNT	Microcrystalline	458	24	5.0	n.d.[k]	36	4	40	[40]
34	2.0 wt% Pt/BP2000	Ball-milled	463	24	5.0	82	49	9	58	[170]
35	5.0 wt% Ru/C, $H_4SiW_{12}O_{40}$	Ball-milled	463	1	9.5	100	n.d.[k]	n.d.[k]	85	[176]
36	5.0 wt% Ru/C, $H_4SiW_{12}O_{40}$	Microcrystalline	463	1	9.5	77	n.d.[k]	n.d.[k]	36	[176]
37	16 wt% Ni_2P/AC	Microcrystalline	498	1.5	6.0	100	48	5	53	[178]
38	$Ni_{12}P_5$/AC[d]	Ball-milled	503	0.7	5.0	92	62	5	67	[179]
39	3.0 wt% Ni/CNF	Ball-milled	463	24	6.0	92	50	6	57	[180]
40	3.0 wt% Ni/CNF	Microcrystalline	483	24	6.0	87	30	5	35	[180]
41	7.5 wt% Ni/oxidized CNF	Ball-milled	463	24	6.0	93	64	7	71	[181]
42	4.0 wt% Ir-4.0 wt% Ni/MC	Microcrystalline	518	0.5	6.0	100	47	12	58	[182]
43	Pt/Ni/mesoporous Al_2O_3[e]	Microcrystalline	473	6	5.0	49	26	6	32	[183]
44	Pt/Ni/BEA[f]	Microcrystalline	473	6	5.0	51	34	3	37	[183]
45	2.0 wt% Ru/AC	Ball-milled	463	18	0.8	83	30	8	38	[184]
46	2.0 wt% Ru/AC	Mix-milled[i]	463	3	0.9	89	58	9	68	[185]
47	Pt, $H_4SiW_{12}O_{40}^g$	Ball-milled	333	24	0.7	n.d.[k]	54	n.d.[k]	n.d.[k]	[112]

(continued)

Table 1.3 (continued)

Entry	Catalyst	Pretreatment of cellulose	T/K	Time/h	$P(H_2)$/MPa	Conv./%	Yield/%			References
							Sor[a]	Man[b]	Total[c]	
48	2.0 wt% Ru/AC	Ball-milled	463	18	0.0[j]	74	34	9	43	[184]
49	2.0 wt% Ru/CMK-3	Ball-milled	463	18	0.0[j]	81	36	9	45	[184]

[a]Sorbitol
[b]Mannitol
[c]Total yield of sorbitol and mannitol
[d]Ni and P loadings were 10 and 2.6 wt%, respectively
[e]Ni/Pt ratio = 22
[f]Ni/Pt ratio = 22. Si/Al ratio of BEA = 75
[g]Pt amount was 24 mol% to cellulose and the concentration of $H_4SiW_{12}O_{40}$ was 0.70 M
[h]Substrate was treated by sulfurous acid to extract lignin, and resulting powder was used as a substrate
[i]Cellulose and catalyst were ball-milled together
[j]2-Propanol was used instead of H_2
[k]No data
[l]Monoanhydrides of sorbitol

catalysts, has been observed in ICP and transmission electron microscope (TEM) analyses. A CNT support improved the yield of sugar alcohols to 73 % when hydrolytic hydrogenating H_3PO_4-pretreated cellulose (entry 32) [40]. This Ru/CNT catalyst also produced sugar alcohols in 40 % yield from microcrystalline cellulose (entry 33). A carbon black named BP2000 was also used as a support, and Pt/BP2000 afforded sugar alcohols in 58 % yield (entry 34) [170]. Both CNT and BP2000 supported metal catalysts showed good reusability. In the hydrolytic hydrogenation of cellulose to sorbitol, the rate determining step is the hydrolysis of cellulose to glucose [170]. To accelerate this rate determining step, $H_4SiW_{12}O_{40}$ was used with Ru/C [176]. This strategy gave 85 % yield of sugar alcohols from ball-milled cellulose (entry 35) and 36 % yield from crystalline cellulose (entry 36). Such high yielding and selective production of sugar alcohols from cellulose is surprising, as side reactions decrease yield and selectivity in typical catalytic reactions. Indeed, similar catalytic systems suffer from side reactions such as further transformation of sorbitol at a higher concentration of acids for a longer reaction time, in which C2–C6 polyols, e.g., erythritol, xylitol, sorbitan, and isosorbide, were obtained as by-products [177].

Supported Ni catalysts have been developed as alternatives to precious metal catalysts to reduce cost. Zhang et al. and Fukuoka et al. impregnated nickel phosphides such as Ni_2P and $Ni_{12}P_5$ on AC [178, 179]. The former species produced sorbitol (48 % yield) and mannitol (5 %) from microcrystalline cellulose (entry 37), and the latter produced sorbitol (62 %) and mannitol (5 %) from ball-milled cellulose (entry 38). However, Ni sintering and P leaching readily occurred in hot compressed water, causing deactivation of these catalysts. In contrast, Ni species on carbon nanofibers (CNFs), which was prepared by chemical vapor deposition (CVD) using methane on Ni/γ-Al_2O_3, was reported to be a reusable catalyst up to three times in water for 24 h even at 483 K (entries 39 and 40) [180]. Later, the catalytic activity of this catalyst was improved by the following procedure [181]. Ni/CNF was dispersed in concentrated HNO_3 at 383 K to oxidize CNF and remove Ni species, and then Ni was again impregnated on the oxidized CNF. Owing to larger amount of OFGs of the oxidized CNF, the dispersion of Ni nanoparticles was improved. The resulting Ni/oxidized CNF catalyst showed higher activity (sorbitol yield 64 %, mannitol 7 %, entry 41) than the original Ni/CNF (sorbitol 50 %, mannitol 6 %, entry 39). The second metals enabled Ni catalysts to improve their functions, and bimetallic catalysts such as Ir–Ni and Pt–Ni were reported for the hydrolytic hydrogenation of cellulose to hexitols [182, 183]. Ir–Ni/MC afforded 58 % yield of sugar alcohols in four cycles of repeated reactions (entry 42) [182]; Ir species thus improved the thermal stability and catalytic activity. The dope of Pt enhanced catalytic performance (entries 43 and 44) compared to only Ni-impregnated catalysts [183]. Nevertheless, the use of precious metals as dopants conflict with the original concept to develop Ni catalysts.

All the hydrolytic hydrogenation reactions described above require H_2 pressure higher than 2 MPa. Fukuoka et al. demonstrated the reaction using a highly dispersed Ru/AC under H_2 as low as 0.8 MPa [184]. In this case, sorbitol and mannitol yields were 30 and 8 %, respectively (entry 45). The yields of sorbitol and mannitol

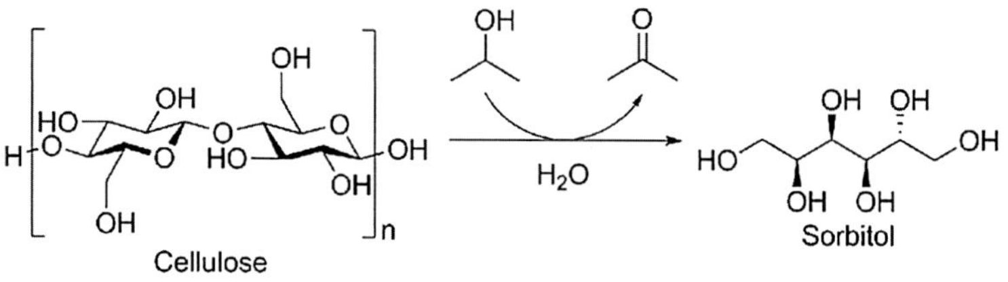

Fig. 1.20 Hydrolytic transfer-hydrogenation of cellulose to sorbitol using 2-propanol as a hydrogen source

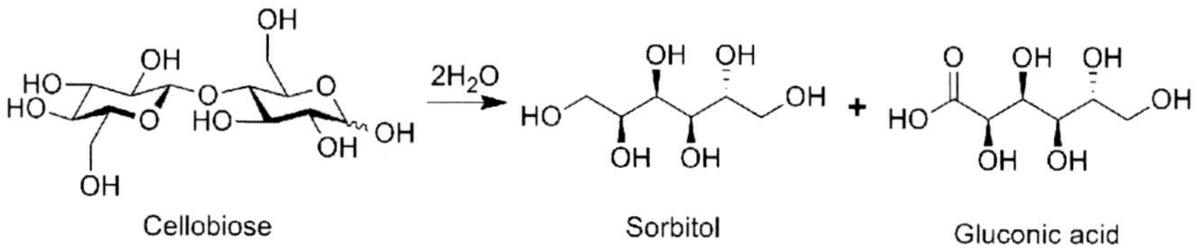

Fig. 1.21 Simultaneous production of sorbitol and gluconic acid from cellobiose

increased up to 58 and 9 % under 0.7–0.9 MPa of H_2 pressure by optimizing pretreatment and reaction conditions (entry 46) [185]. The combination of a high-loaded Pt (24 mol% to cellulose) catalyst and concentrated $H_4SiW_{12}O_{40}$ (0.70 M) also enabled to reduce H_2 pressure to 0.7 MPa (entry 47) [112].

The hydrolytic transfer-hydrogenation of cellulose to sorbitol (Fig. 1.20) was promoted by Ru/AC catalyst (entries 48 and 49) [184]. Other supports (TiO_2, ZrO_2, and Al_2O_3) or metals (Rh, Ir, Pd, Pt, and Au) were inactive for this reaction. An active hydrogen species is generated from 2-propanol over Ru nanoparticles during the reaction, and thus the addition of H_2 is not necessary in this catalytic system.

Another type of reaction to produce sorbitol from β-glucan has been reported. Under Ar atmosphere, disproportionation including both reduction and oxidation of reducing ends in glucan molecules takes place over Ru/AC, resulting in the simultaneous production of sorbitol and gluconic acid from cellobiose (Fig. 1.21) [186]. In addition to sorbitol, gluconic acid is a useful chemical as food additive and medicine [68, 187]. In this reaction, hydrogen species is provided by oxidation of a reducing end, and thus hydrogen sources such as H_2 gas and 2-propanol are not required.

1.3 Chitin

1.3.1 Property of Chitin

Chitin is the most abundant nitrogen-containing biomass and its annual production by crustaceans in oceans is 10^{11} tons [188]. The structure of the chitin molecule is the same as that of cellulose except for a functional group at C2 position (Fig. 1.22);

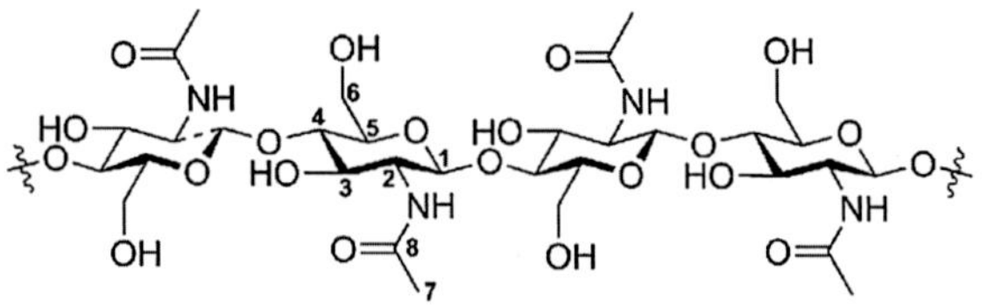

Fig. 1.22 Structure of chitin

C2 in chitin coordinates to an acetamido group instead of a hydroxyl group. A monomeric unit possessing the acetamido group is called *N*-acetylglucosamine (denoted as GlcNAc in this thesis. Another abbreviation NAG is widely used as well). Another nitrogen-containing polymer called chitosan is also known as one of the marine biomass resources, and a building block of chitosan is glucosamine (GlcN), which has an amino group at C2 position. Chitin has robust crystalline structures formed by inter- and intramolecular hydrogen bond network similar to cellulose. There are three crystalline types, i.e., α-, β-, and γ-chitin, distinguished by packing manner of chitin strands, and the type depends on the origin of chitin. The molecules in α-chitin are assembled as antiparallel chain packing and those in β-chitin are done as parallel packing [189]. γ-Chitin includes both packing senses, in which two chains run in the same direction and one chain does in the opposite direction. Chitin itself is utilized in several areas, e.g., biocompatible materials [190], column packing materials for affinity chromatography [191], catalyst supports owing to its high stability [192], and source of nitrogen-doped carbons [193].

In order to disrupt the crystalline structure and improve reactivity of chitin, pretreatment has been conducted for chitin as well as that for cellulose described in Sect. 1.2.2. Totani et al. reported that converge milling changed crystalline chitin to an amorphous form for 30 min, decreasing *CrI* from 91 to 40 %, calculated by Eq. 1.4 [194] and XRD patterns [195].

$$CrI_{XRD} = \frac{I_{20} - I_{16}}{I_{20}} \times 100 \qquad (1.4)$$

where I_{20} and I_{16} are intensities at 20° and 16° in XRD pattern, respectively.

Osada et al. applied sub- and supercritical water to decomposition of the crystalline structure of chitin [196]. After supercritical water treatment at 673 K for 2 min, *CrI* decreased from 90 to 50 %. The *d*-spacing determined from XRD slightly increased by the treatment, implying that intermolecular hydrogen bonds of chitin were weakened and/or divided. Both the decrease of *CrI* and increase of *d*-spacing would lead to easy access of catalysts to reaction sites of chitin. They also demonstrated the combination of supercritical water treatment and converge milling [197]. The value of *CrI* was 26 % in this case, clearly indicating that the combination of both methods is more effective than each individual method.

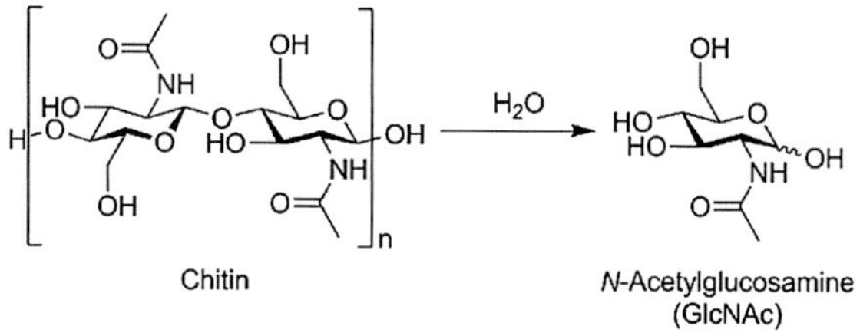

Fig. 1.23 Hydrolysis of chitin to GlcNAc

1.3.2   Transformation of Chitin

Hydrolysis of chitin to GlcNAc (Fig. 1.23) is an attractive reaction since GlcNAc is a potential substrate as dietary supplements [198], cosmetics [9], and medicines against arthritic and inflammatory bowel diseases [199]. It is notable that deacetylation of GlcNAc to GlcN causes drop of physiological activity and usability [200, 201]; in other words, GlcNAc is superior to GlcN in applications. Hence, the production of GlcNAc from chitin with retention of *N*-acetyl group is favorable.

Chitinase enzymes, i.e., endo- and exochitinase, depolymerize crystalline chitin to GlcNAc in a high yield of 77 % [202]. Although enzymes work under ambient conditions at 290 K, it takes 10 days to complete the reaction. Osada et al. drastically reduced the reaction time to 24 h by the supercritical water and converge milling pretreatment of the substrate as described in Sect. 1.3.1 [197]. The resulting product was chitobiose, a dimer of GlcNAc, in 93 % yield due to the use of only endochitinase. Exochitinase or chitobiase is necessary for further depolymerization of chitobiose to GlcNAc. Chitinase enzymes recognize and selectively hydrolyze glycosidic bonds in chitin under mild conditions, in which chitin and hydrolysates do not undergo deacetylation. Consequently, *N*-acetyl groups are completely retained in enzymatic reactions.

Hydrolysis of chitin in a concentrated HCl solution (15–36 %) has also afforded GlcNAc in ca. 65 % yield with preserving *N*-acetyl groups [203]. Intriguingly, the rate constant for hydrolysis of glycosidic bonds is >100 times higher than that of *N*-acetyl groups at 298–308 K in a 12 M of HCl solution. This large difference in reaction rates contributes to the high-yielding production of GlcNAc from chitin with suppressing deacetylation. However, this system suffers from several drawbacks, including corrosive property of concentrated HCl and a huge amount of acidic waste [9]. In a diluted HCl solution (3 M), the ratio of the rate constants (depolymerization/deacetylation) is only ca. 2 at 298–308 K, and thus depolymerization and deacetylation simultaneously happen to provide GlcN as a major product [203].

GlcNAc can be derivatized to nitrogen-containing compounds (Fig. 1.24) such as 1-*O*-methyl-*N*-acetylglucosamine (MeGlcNAc) [204], 3-acetamido-5-acetylfuran (3A5AF) [205, 206], 2-acetamido-2,3-dideoxy-erythro-hex-2-enofuranose (Chromogen I) [207, 208], 3-acetamido-5-(1,2-dihydroxyethyl)furan (Chromogen III) [207, 208],

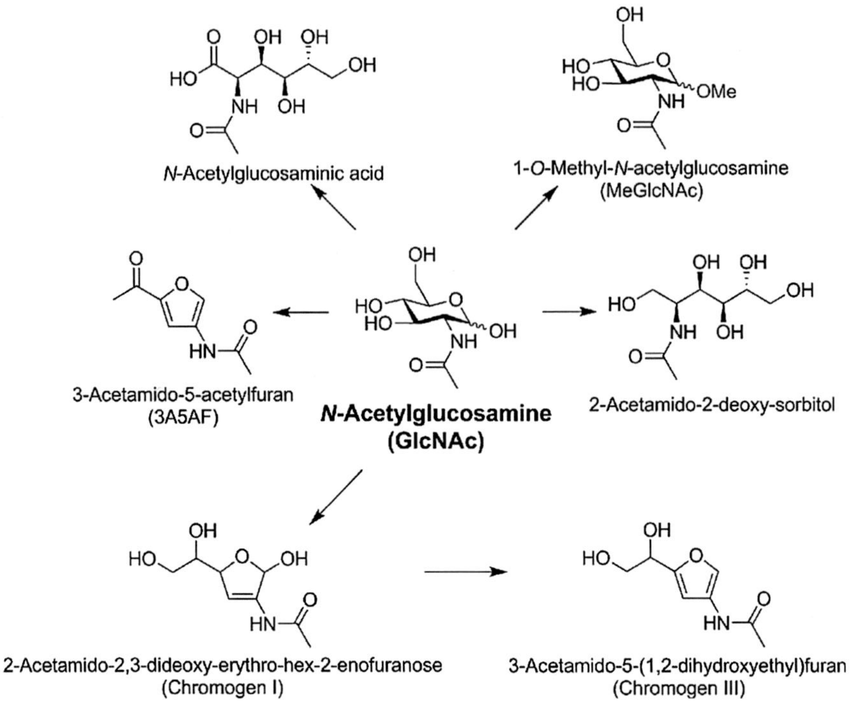

Fig. 1.24 Derivatization of GlcNAc

N-acetylglucosaminic acid [209], and 2-acetamido-2-deoxy-sorbitol [210], which are biologically-active agents and potential precursors to medicines and polymers. Recently, Yan and co-workers reported the direct production of 3A5AF from chitin by boric acid [211]. Boric acid gave 3.6 % yield of 3A5AF from crystalline chitin, and the additives of LiCl and HCl further increased the yield to 6.2 %. They proposed that boric acid formed a bidentate complex with GlcNAc during the reaction. As a result, GlcNAc easily underwent both activation and dehydration to form 3A5AF. Meanwhile, Cl⁻ species possibly disrupted hydrogen bonds of chitin and improved its reactivity; this function of Cl⁻ for chitin is the same as that for cellulose (see Sect. 1.2.2). Yan and co-workers also demonstrated the degradation of chitin in ethylene glycol in the presence of H_2SO_4 [212]. Both glycosidic bonds and *N*-acetamido groups were cleaved under their conditions, and the major product was 1-*O*-(2-hydroxyethyl)-2-amino-2-deoxyglucopyranoside (HADP, ca. 25 %, Fig. 1.25). The *N*-acetamido-groups-retaining product, 1-*O*-(2-hydroxyethyl)-2-acetamido-2-deoxyglucopyranoside (HAADP), was obtained in only <10 % yield due to deacetylation.

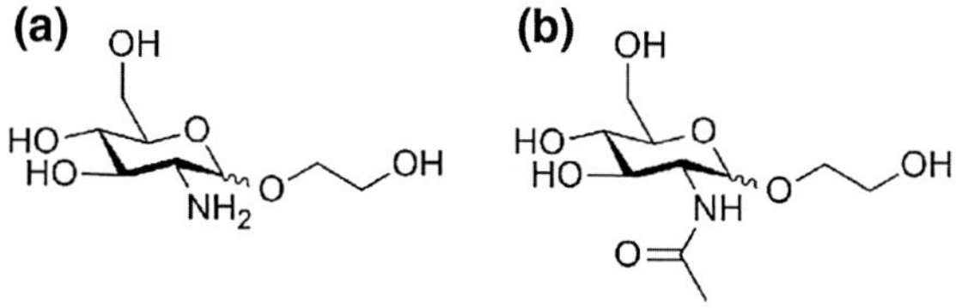

Fig. 1.25 Structures of **a** HADP and **b** HAADP

1.4 Objectives of This Thesis Work

As described above, the hydrolysis of cellulose to produce glucose is one of the main target reactions in biorefinery. H_2SO_4 and cellulase enzymes have been used for the depolymerization even in commercial scales, but their complete recovery requires unprofitable processes. Hence, to date, heterogeneous catalysts have attracted great interest due to their ease of separation and reuse. In this area, most researchers focus on strong acid catalysts, namely sulfonated materials, and indeed they efficiently hydrolyze *pure* cellulose to glucose in 75–92 % yields as described in Sect. 1.2.5; however, their durability and S/C ratio are low in some cases. Additionally, for practical use, catalysts need to preserve their catalytic activity even in the presence of salts derived from raw biomass, but strong acids are easily deactivated by ion exchange due to their low pK_a [154, 213]. Interestingly, it has been found that carbon materials even without sulfonation promote the hydrolysis of cellulose, and weakly acidic active sites are expected to resist ion exchange with salts. Hence, one of the objectives of this work is depolymerization of cellulose and raw biomass to sugars using carbon materials without strong acid sites in high efficiency. Meanwhile, the structure–activity correlation of carbons has been studied by physicochemical characterization, thermodynamics, kinetics, model reactions, and computations.

Next, the author focuses on the efficient transformation of chitin with retention of *N*-acetyl groups, which is a grand challenge in marine biomass refinery. The depolymerization of chitin without deacetylation has been achieved only using chitinase enzymes and concentrated HCl, and these systems suffer from severe issues such as long reaction times, equipment corrosion, and a huge amount of acidic waste. Herein, the author has developed a new catalytic system to overcome these drawbacks and depolymerize chitin to monomers with preserving *N*-acetyl groups.

The third target is selective production of 1,4-sorbitan from sorbitol, which is in high demand due to its wide applications. The selective production of 1,4-sorbitan has been hampered by successive dehydration of 1,4-sorbitan. Hence, the author has investigated the dehydration of sorbitol to provide 1,4-sorbitan with good yield and selectivity by using both homo- and heterogeneous acids.

1.5   Outlines of This Thesis

1.5.1   Chapter 1: General Introduction

The importance of biomass utilization in the current world situations is described. Subsequently, the chemical transformations of cellulose and chitin using catalysts and their remaining issues are discussed. Then, the goal of this thesis work is stated.

1.5.2   Chapter 2: Hydrolysis of Cellulose to Glucose Using Carbon Catalysts

The objective in this chapter is high-yielding production of glucose from cellulose using carbon catalysts. Alkali-activated carbons show significantly high catalytic activity for the hydrolysis of cellulose. A new pretreatment method, named mix-milling, is developed to drastically accelerate the hydrolysis of cellulose by the carbons, resulting in one of the highest yields of glucose in any methods. The effects of mix-milling on cellulose hydrolysis are investigated by characterization, model reactions, and kinetics.

1.5.3   Chapter 3: Mechanistic Study of Cellulose Hydrolysis by Carbon Catalysts

The catalysis of carbon materials for hydrolysis of glycosidic bonds is scrutinized in this chapter. The author proposes active sites on carbons and their catalysis from physicochemical characterization of carbons, model reactions, kinetics, thermodynamics, and computations.

1.5.4   Chapter 4: Catalytic Depolymerization of Chitin to N-Acetylated Monomers

The catalytic two-step depolymerization of chitin to monomers with retention of N-acetyl groups has been developed. A combination of mechanocatalytic hydrolysis and thermocatalytic solvolysis, hydrolysis and methanolysis in this work, provides N-acetylated monomers in good yields.

1.5.5 Chapter 5: Acid-Catalyzed Dehydration of Sorbitol to 1,4-Sorbitan

The sorbitol dehydration catalyzed by solid acids as well as H_2SO_4 is investigated in detail by identification of reaction products and kinetics. The results represented here provide a blueprint for designing new solid catalysts for the selective synthesis of anhydropolyols.

1.5.6 Chapter 6: General Conclusions

This chapter summarizes all of the results and discussion and shows their significances in both the scientific and social sense.

References

1. Goldemberg J (2007) Ethanol for a sustainable energy future. Science 315(5813):808–810
2. Román-Leshkov Y, Barrett CJ, Liu ZY, Dumesic JA (2007) Production of dimethylfuran for liquid fuels from biomass-derived carbohydrates. Nature 447(7147):982–985
3. Chheda JN, Huber GW, Dumesic JA (2007) Liquid-phase catalytic processing of biomass-derived oxygenated hydrocarbons to fuels and chemicals. Angew Chem Int Ed 46 (38):7164–7183
4. Corma A, Iborra S, Velty A (2007) Chemical routes for the transformation of biomass into chemicals. Chem Rev 107(6):2411–2502
5. Himmel ME, Ding S-Y, Johnson DK, Adney WS, Nimlos MR, Brady JW, Foust TD (2007) Biomass recalcitrance: engineering plants and enzymes for biofuels production. Science 315 (5813):804–807
6. Gallezot P (2008) Catalytic conversion of biomass: challenges and issues. ChemSusChem 1 (8–9):734–737
7. Fukuoka A, Dhepe PL (2009) Sustainable green catalysis by supported metal nanoparticles. Chem Rec 9(4):224–235
8. Kobayashi H, Ohta H, Fukuoka A (2012) Conversion of lignocellulose into renewable chemicals by heterogeneous catalysis. Catal Sci Technol 2(5):869–883
9. Chen J-K, Shen C-R, Liu C-L (2010) N-Acetylglucosamine: production and applications. Mar Drugs 8(9):2493–2516
10. Chen X, Yan N (2014) Novel catalytic systems to convert chitin and lignin into valuable chemicals. Catal Surv Asia 18(4):164–176
11. McKendry P (2002) Energy production from biomass (part 1): overview of biomass. Bioresour Technol 83(1):37–46
12. Mosier N, Wyman C, Dale B, Elander R, Lee YY, Holtzapple M, Ladisch M (2005) Features of promising technologies for pretreatment of lignocellulosic biomass. Bioresour Technol 96 (6):673–686
13. Walther T, Hensirisak P, Agblevor FA (2001) The influence of aeration and hemicellulosic sugars on xylitol production by Candida tropicalis. Bioresour Technol 76(3):213–220
14. Palm M, Zacchi G (2004) Separation of hemicellulosic oligomers from steam-treated spruce wood using gel filtration. Sep Purif Technol 36(3):191–201

15. Ralph J, Lundquist K, Brunow G, Lu F, Kim H, Schatz PF, Marita JM, Hatfield RD, Ralph SA, Christensen JH, Boerjan W (2004) Lignins: natural polymers from oxidative coupling of 4-hydroxyphenyl-propanoids. Phytochem Rev 3(1):29–60

16. Meier D, Faix O (1999) State of the art of applied fast pyrolysis of lignocellulosic materials —a review. Bioresour Technol 68(1):71–77

17. Mohan D, Pittman CU, Steele PH (2006) Pyrolysis of wood/biomass for bio-oil: a critical review. Energy Fuels 20(3):848–889

18. Pandey MP, Kim CS (2011) Lignin depolymerization and conversion: a review of thermochemical methods. Chem Eng Technol 34(1):29–41

19. Zugenmaier P (2001) Conformation and packing of various crystalline cellulose fibers. Prog Polym Sci 26(9):1341–1417

20. Rinaldi R, Schüth F (2009) Acid hydrolysis of cellulose as the entry point into biorefinery schemes. ChemSusChem 2(12):1096–1107

21. Sugiyama J, Vuong R, Chanzy H (1991) Electron diffraction study on the two crystalline phases occurring in native cellulose from an algal cell wall. Macromolecules 24(14):4168–4175

22. Langan P, Nishiyama Y, Chanzy H (2001) X-ray structure of mercerized cellulose II at 1 Å resolution. Biomacromolecules 2(2):410–416

23. Hayashi J, Sufoka A, Ohkita J, Watanabe S (1975) The confirmation of existences of cellulose III$_I$, III$_{II}$, IV$_I$, and IV$_{II}$ by the X-ray method. J Polym Sci Polym Lett Ed 13(1):23–27

24. Drisch N, Herrbach P (1955) Process for the production of rayon products. US Patent 2,705,184

25. Kawano S, Tajima K, Uemori Y, Yamashita H, Erata T, Munekata M, Takai M (2002) Cloning of cellulose synthesis related genes from *Acetobacter xylinum* ATCC23769 and ATCC53582: comparison of cellulose synthetic ability between strains. DNA Res 9(5):149–156

26. Reddy KR, Kumar NS, Reddy PS, Sreedhar B, Kantam ML (2006) Cellulose supported palladium(0) catalyst for Heck and Sonogashira coupling reactions. J Mol Catal A Chem 252(1):12–16

27. Cirtiu CM, Dunlop-Brière AF, Moores A (2011) Cellulose nanocrystallites as an efficient support for nanoparticles of palladium: application for catalytic hydrogenation and Heck coupling under mild conditions. Green Chem 13(2):288–291

28. Yano H, Nakahara S (2008) High strength material using cellulose microfibrils. US Patent 7,378,149

29. Kondo T, Morita M, Hayakawa K, Onda Y (2008) Wet pulverizing of polysaccharides. US Patent 7,357,339

30. Zhao H, Kwak JH, Wang Y, Franz JA, White JM, Holladay JE (2006) Effects of crystallinity on dilute acid hydrolysis of cellulose by cellulose ball-milling study. Energy Fuels 20 (2):807–811

31. Segal L, Creely JJ, Martin AE Jr, Conrad CM (1959) An empirical method for estimating the degree of crystallinity of native cellulose using the X-ray diffractometer. Text Res J 29 (10):786–794

32. Zhang Y-HP, Lynd LR (2004) Toward an aggregated understanding of enzymatic hydrolysis of cellulose: noncomplexed cellulase systems. Biotechnol Bioeng 88(7):797–824

33. Wikberg H, Maunu SL (2004) Characterisation of thermally modified hard- and softwoods by ^{13}C CPMAS NMR. Carbohydr Polym 58(4):461–466

34. Park S, Johnson DK, Ishizawa CI, Parilla PA, Davis MF (2009) Measuring the crystallinity index of cellulose by solid state ^{13}C nuclear magnetic resonance. Cellulose 16(4):641–647

35. Mazeau K, Heux L (2003) Molecular dynamics simulations of bulk native crystalline and amorphous structures of cellulose. J Phys Chem B 107(10):2394–2403

36. Fan LT, Lee Y-H, Beardmore DR (1981) The influence of major structural features of cellulose on rate of enzymatic hydrolysis. Biotechnol Bioeng 23(2):419–424

37. Yabushita M, Kobayashi H, Fukuoka A (2014) Catalytic transformation of cellulose into platform chemicals. Appl Catal B Environ 145:1–9
38. Suzuki T, Nakagami H (1999) Effect of crystallinity of microcrystalline cellulose on the compactability and dissolution of tablets. Eur J Pharm Biopharm 47(3):225–230
39. Avolio R, Bonadies I, Capitani D, Errico ME, Gentile G, Avella M (2012) A multitechnique approach to assess the effect of ball milling on cellulose. Carbohydr Polym 87(1):265–273
40. Deng W, Tan X, Fang W, Zhang Q, Wang Y (2009) Conversion of cellulose into sorbitol over carbon nanotube-supported ruthenium catalyst. Catal Lett 133(1):167–174
41. Charmot A, Katz A (2010) Unexpected phosphate salt-catalyzed hydrolysis of glycosidic bonds in model disaccharides: cellobiose and maltose. J Catal 276(1):1–5
42. Zheng Y, Lin H-M, Tsao GT (1998) Pretreatment for cellulose hydrolysis by carbon dioxide explosion. Biotechnol Prog 14(6):890–896
43. Alvira P, Tomás-Pejó E, Ballesteros M, Negro MJ (2010) Pretreatment technologies for an efficient bioethanol production process based on enzymatic hydrolysis: a review. Bioresour Technol 101(13):4851–4861
44. Chang VS, Holtzapple MT (2000) Fundamental factors affecting biomass enzymatic reactivity. In: Finkelstein M, Davison BH (eds) Twenty-first symposium on biotechnology for fuels and chemicals. Humana Press, New York, pp 5–37
45. Zhao H, Kwak JH, Wang Y, Franz JA, White JM, Holladay JE (2007) Interactions between cellulose and N-methylmorpholine-N-oxide. Carbohydr Polym 67(1):97–103
46. Strachan J (1938) Solubility of cellulose in water. Nature 141:332–333
47. Fang Z, Koziński JA (2000) Phase behavior and combustion of hydrocarbon-contaminated sludge in supercritical water at pressures up to 822 MPa and temperatures up to 535 ⁰C. Proc Combust Inst 28(2):2717–2725
48. McCormick CL, Callais PA, Hutchinson BH Jr (1985) Solution studies of cellulose in lithium chloride and N,N-dimethylacetamide. Macromolecules 18(12):2394–2401
49. Röder T, Morgenstern B, Schelosky N, Glatter O (2001) Solutions of cellulose in N,N-dimethylacetamide/lithium chloride studied by light scattering methods. Polymer 42 (16):6765–6773
50. Burchard W, Habermann N, Klüfers P, Seger B, Wilhelm U (1994) Cellulose in Schweizer's reagent: a stable, polymeric metal complex with high chain stiffness. Angew Chem Int Ed 33 (8):884–887
51. Kauffman GB (1993) Rayon: the first semi-synthetic fiber product. J Chem Educ 70(11):887
52. Swatloski RP, Spear SK, Holbrey JD, Rogers RD (2002) Dissolution of cellose with ionic liquids. J Am Chem Soc 124(18):4974–4975
53. Brennecke JF, Maginn EJ (2001) Ionic liquids: innovative fluids for chemical processing. AIChE J 47(11):2384–2389
54. Lamed R, Zeikus JG (1980) Glucose fermentation pathway of Thermoanaerobium brockii. J Bacteriol 141(3):1251–1257
55. Lin Y, Tanaka S (2006) Ethanol fermentation from biomass resources: current state and prospects. Appl Microbiol Biotechnol 69(6):627–642
56. Sun Y, Cheng J (2002) Hydrolysis of lignocellulosic materials for ethanol production: a review. Bioresour Technol 83(1):1–11
57. Zhao G, Zheng M, Zhang J, Wang A, Zhang T (2013) Catalytic conversion of concentrated glucose to ethylene glycol with semicontinuous reaction system. Ind Eng Chem Res 52 (28):9566–9572
58. Matsumoto K, Kobayashi H, Ikeda K, Komanoya T, Fukuoka A, Taguchi S (2011) Chemo-microbial conversion of cellulose into polyhydroxybutyrate through ruthenium-catalyzed hydrolysis of cellulose into glucose. Bioresour Technol 102(3):3564–3567
59. Nduko JM, Suzuki W, Matsumoto K, Kobayashi H, Ooi T, Fukuoka A, Taguchi S (2012) Polyhydroxyalkanoates production from cellulose hydrolysate in Escherichia coli LS5218 with superior resistance to 5-hydroxymethylfurfural. J Biosci Bioeng 113(1):70–72

60. Gallezot P, Nicolaus N, Fleche G, Fuertes P, Perrard A (1998) Glucose hydrogenation on ruthenium catalysts in a trickle-bed reactor. J Catal 180(1):51–55

61. Li H, Wang W, Deng JF (2000) Glucose hydrogenation to sorbitol over a skeletal Ni-P amorphous alloy catalyst (Raney Ni-P). J Catal 191(1):257–260

62. Li H, Li H, Deng J-F (2002) Glucose hydrogenation over Ni–B/SiO$_2$ amorphous alloy catalyst and the promoting effect of metal dopants. Catal Today 74(1):53–63

63. Hoffer BW, Crezee E, Devred F, Mooijman PRM, Sloof WG, Kooyman PJ, van Langeveld AD, Kapteijn F, Moulijn JA (2003) The role of the active phase of Raney-type Ni catalysts in the selective hydrogenation of D-glucose to D-sorbitol. Appl Catal A Gen 253 (2):437–452

64. Werpy T, Petersen G, Aden A, Bozell J, Holladay J, White J, Manheim A, Eliot D, Lasure L, Jones S, Gerber M, Ibsen K, Lumberg L, Kelley S. Top value added chemicals from biomass. Volume I: results of screening for potential candidates from sugars and synthesis gas. (Online) http://www.dtic.mil/cgi-bin/GetTRDoc?Location=U2&doc=GetTRDoc.pdf&AD= ADA436528 Accessed 31 Oct 2015

65. Zhang J, Li J-b WuS-B, Liu Y (2013) Advances in the catalytic production and utilization of sorbitol. Ind Eng Chem Res 52(34):11799–11815

66. Chen K, Tamura M, Yuan Z, Nakagawa Y, Tomishige K (2013) One-pot conversion of sugar and sugar polyols to *n*-alkanes without C–C dissociation over the Ir-ReO$_x$/SiO$_2$ catalyst combined with H-ZSM-5. ChemSusChem 6(4):613–621

67. Shiramizu M, Toste FD (2012) Deoxygenation of biomass-derived feedstocks: oxorhenium-catalyzed deoxydehydration of sugars and sugar alcohols. Angew Chem Int Ed 51(32):8082–8086

68. Kobayashi H, Fukuoka A (2013) Synthesis and utilisation of sugar compounds derived from lignocellulosic biomass. Green Chem 15(7):1740–1763

69. Yoshioka T, Sternberg B, Florence AT (1994) Preparation and properties of vesicles (niosomes) of sorbitan monoesters (Span 20, 40, 60 and 80) and a sorbitan triester (Span 85). Int J Pharm 105(1):1–6

70. Fenouillot F, Rousseau A, Colomines G, Saint-Loup R, Pascault J-P (2010) Polymers from renewable 1,4:3,6-dianhydrohexitols (isosorbide, isomannide and isoidide): a review. Prog Polym Sci 35(5):578–622

71. Fuji M, Akita M, Tanaka T (2010) Polycarbonate copolymer and method of producing the same. US Patent 2010/0190953

72. Jansen BJP, Kamps JH, Kung E, Looij H, Prada L, Steendam WJD (2010) Isosorbide-based polycarbonates, method of making, and articles formed therefrom. US Patent 7,666,972

73. Oda A, Kitazono E, Miyake T, Kinoshita M, Saito M (2011) Polycarbonate containing plant-derived component and process for the preparation thereof. US Patent 8,017,722

74. Gohil RM (2009) Properties and strain hardening character of polyethylene terephthalate containing isosorbide. Polym Eng Sci 49(3):544–553

75. Andrews MA, Bhatia KK, Fagan PJ (2004) Process for the manufacture of anhydro sugar alcohols with the assistance of a gas purge. US Patent 6,689,892

76. Sun P, Yu DH, Hu Y, Tang ZC, Xia JJ, Li H, Huang H (2011) H$_3$PW$_{12}$O$_{40}$/SiO$_2$ for sorbitol dehydration to isosorbide: high efficient and reusable solid acid catalyst. Korean J Chem Eng 28(1):99–105

77. Gu M, Yu D, Zhang H, Sun P, Huang H (2009) Metal (IV) phosphates as solid catalysts for selective dehydration of sorbitol to isosorbide. Catal Lett 133(1):214–220

78. Xia J, Yu D, Hu Y, Zou B, Sun P, Li H, Huang H (2011) Sulfated copper oxide: an efficient catalyst for dehydration of sorbitol to isosorbide. Catal Commun 12(6):544–547

79. Montassier C, Ménézo J, Moukolo J, Naja J, Hoang LC, Barbier J, Boitiaux JP (1991) Polyol conversions into furanic derivatives on bimetallic catalysts: Cu–Ru, Cu–Pt and Ru–Cu. J Mol Catal 70(1):65–84

80. Moore KM, Sanborn AJ, Bloom P (2008) Process for the production of anhydrosugar alcohols. US Patent 7,439,352

81. Khan NA, Mishra DK, Ahmed I, Yoon JW, Hwang J-S, Jhung SH (2013) Liquid-phase dehydration of sorbitol to isosorbide using sulfated zirconia as a solid acid catalyst. Appl Catal A Gen 452:34–38

82. Ahmed I, Khan NA, Mishra DK, Lee JS, Hwang J-S, Jhung SH (2013) Liquid-phase dehydration of sorbitol to isosorbide using sulfated titania as a solid acid catalyst. Chem Eng Sci 93:91–95

83. Sanborn AJ (2008) Process for the production of anhydrosugar alcohols. US Patent 7,420,067

84. Liu A, Luckett CC (2011) Sorbitol conversion process. US Patent 7,982,059

85. Kobayashi H, Yokoyama H, Feng B, Fukuoka A (2015) Dehydration of sorbitol to isosorbide over H-beta zeolites with high Si/Al ratios. Green Chem 17(5):2732–2735

86. Otomo R, Yokoi T, Tatsumi T (2015) Synthesis of isosorbide from sorbitol in water over high-silica aluminosilicate zeolites. Appl Catal A Gen 505:28–35

87. Morita Y, Furusato S, Takagaki A, Hayashi S, Kikuchi R, Oyama ST (2014) Intercalation-controlled cyclodehydration of sorbitol in water over layered-niobium-molybdate solid acid. ChemSusChem 7(3):748–752

88. Teong SP, Yi G, Zhang Y (2014) Hydroxymethylfurfural production from bioresources: past, present and future. Green Chem 16(4):2015–2026

89. Zhao H, Holladay JE, Brown H, Zhang ZC (2007) Metal chlorides in ionic liquid solvents convert sugars to 5-hydroxymethylfurfural. Science 316(5831):1597–1600

90. Román-Leshkov Y, Moliner M, Labinger JA, Davis ME (2010) Mechanism of glucose isomerization using a solid Lewis acid catalyst in water. Angew Chem Int Ed 49(47):8954–8957

91. Akien GR, Qi L, Horváth IT (2012) Molecular mapping of the acid catalysed dehydration of fructose. Chem Commun 48(47):5850–5852

92. Nikolla E, Román-Leshkov Y, Moliner M, Davis ME (2011) "One-pot" synthesis of 5-(hydroxymethyl)furfural from carbohydrates using tin-beta zeolite. ACS Catal 1 (4):408–410

93. Yan H, Yang Y, Tong D, Xiang X, Hu C (2009) Catalytic conversion of glucose to 5-hydroxymethylfurfural over SO_4^{2-}/ZrO_2 and SO_4^{2-}/ZrO_2–Al_2O_3 solid acid catalysts. Catal Commun 10(11):1558–1563

94. Otomo R, Tatsumi T, Yokoi T (2015) Beta zeolite: a universally applicable catalyst for the conversion of various types of saccharides into furfurals. Catal Sci Technol 5(8):4001–4007

95. Nakajima K, Noma R, Kitano M, Hara M (2014) Selective glucose transformation by titania as a heterogeneous Lewis acid catalyst. J Mol Catal A Chem 388–389:100–105

96. Shiramizu M, Toste FD (2011) On the Diels-Alder approach to solely biomass-derived polyethylene terephthalate (PET): conversion of 2,5-dimethylfuran and acrolein into p-xylene. Chem Eur J 17(44):12452–12457

97. Williams CL, Chang C-C, Do P, Nikbin N, Caratzoulas S, Vlachos DG, Lobo RF, Fan W, Dauenhauer PJ (2012) Cycloaddition of biomass-derived furans for catalytic production of renewable p-xylene. ACS Catal 2(6):935–939

98. Gorbanev YY, Klitgaard SK, Woodley JM, Christensen CH, Riisager A (2009) Gold-catalyzed aerobic oxidation of 5-hydroxymethylfurfural in water at ambient temperature. ChemSusChem 2(7):672–675

99. Davis SE, Houk LR, Tamargo EC, Datye AK, Davis RJ (2011) Oxidation of 5-hydroxymethylfurfural over supported Pt, Pd and Au catalysts. Catal Today 160(1):55–60

100. Rosatella AA, Simeonov SP, Frade RFM, Afonso CAM (2011) 5-Hydroxymethylfurfural (HMF) as a building block platform: biological properties, synthesis and synthetic applications. Green Chem 13(4):754–793

101. Bozell JJ, Moens L, Elliott DC, Wang Y, Neuenscwander GG, Fitzpatrick SW, Bilski RJ, Jarnefeld JL (2000) Production of levulinic acid and use as a platform chemical for derived products. Resour Conserv Recy 28(3):227–239

102. West RM, Liu ZY, Peter M, Dumesic JA (2008) Liquid alkanes with targeted molecular weights from biomass-derived carbohydrates. ChemSusChem 1(5):417–424

103. Hamelinck CN, van Hooijdonk G, Faaij APC (2005) Ethanol from lignocellulosic biomass: techno-economic performance in short-, middle- and long-term. Biomass Bioenergy 28 (4):384–410

104. Balat M (2011) Production of bioethanol from lignocellulosic materials via the biochemical pathway: a review. Energy Convers Manage 52(2):858–875

105. Faith WL (1945) Development of the Scholler process in the United States. Ind Eng Chem 37 (1):9–11

106. Clausen EC, Gaddy JL (1993) Concentrated sulfuric acid process for converting lignocellulosic materials to sugars. US Patent 5,188,673

107. Hick SM, Griebel C, Restrepo DT, Truitt JH, Buker EJ, Bylda C, Blair RG (2010) Mechanocatalysis for biomass-derived chemicals and fuels. Green Chem 12(3):468–474

108. Meine N, Rinaldi R, Schüth F (2012) Solvent-free catalytic depolymerization of cellulose to water-soluble oligosaccharides. ChemSusChem 5(8):1449–1454

109. Shrotri A, Lambert LK, Tanksale A, Beltramini J (2013) Mechanical depolymerisation of acidulated cellulose: understanding the solubility of high molecular weight oligomers. Green Chem 15(10):2761–2768

110. Hilgert J, Meine N, Rinaldi R, Schüth F (2013) Mechanocatalytic depolymerization of cellulose combined with hydrogenolysis as a highly efficient pathway to sugar alcohols. Energy Environ Sci 6(1):92–96

111. Shrotri A, Kobayashi H, Tanksale A, Fukuoka A, Beltramini J (2014) Transfer hydrogenation of cellulose-based oligomers over carbon-supported ruthenium catalyst in a fixed-bed reactor. ChemCatChem 6(5):1349–1356

112. Ogasawara Y, Itagaki S, Yamaguchi K, Mizuno N (2011) Saccharification of natural lignocellulose biomass and polysaccharides by highly negatively charged heteropolyacids in concentrated aqueous solution. ChemSusChem 4(4):519–525

113. Shimizu K, Furukawa H, Kobayashi N, Itaya Y, Satsuma A (2009) Effects of Brønsted and Lewis acidities on activity and selectivity of heteropolyacid-based catalysts for hydrolysis of cellobiose and cellulose. Green Chem 11(10):1627–1632

114. Walker LP, Wilson DB (1991) Enzymatic hydrolysis of cellulose: an overview. Bioresour Technol 36(1):3–14

115. Medve J, Ståhlberg J, Tjerneld F (1997) Isotherms for adsorption of cellobiohydrolase I and II from *trichoderma reesei* on microcrystalline cellulose. Appl Biochem Biotechnol 66 (1):39–56

116. Davies G, Henrissat B (1995) Structures and mechanisms of glycosyl hydrolases. Structure 3 (9):853–859

117. Divne C, Ståhlberg J, Teeri TT, Jones TA (1998) High-resolution crystal structures reveal how a cellulose chain is bound in the 50 Å long tunnel of cellobiohydrolase I from *Trichoderma reesei*. J Mol Biol 275(2):309–325

118. McCarter JD, Withers GS (1994) Mechanisms of enzymatic glycoside hydrolysis. Curr Opin Struct Biol 4(6):885–892

119. Emert GH, Blotkamp PJ (1980) Method for enzyme reutilization. US Patent 4,220,721

120. Woodward J (1989) Recovery and reuse of cellulase catalyst in an enzymatic cellulose hydrolysis process. US Patent 4,840,904

121. Garcia A III, Oh S, Engler CR (1989) Cellulase immobilization on Fe_3O_4 and characterization. Biotechnol Bioeng 33(3):321–326

122. Hartono SB, Qiao SZ, Liu J, Jack K, Ladewig BP, Hao Z, Lu GQM (2010) Functionalized mesoporous silica with very large pores for cellulase immobilization. J Phys Chem C 114 (18):8353–8362

123. Chang RH-Y, Jang J, Wu KC-W (2011) Cellulase immobilized mesoporous silica nanocatalysts for efficient cellulose-to-glucose conversion. Green Chem 13(10):2844–2850

124. Sasaki M, Kabyemela B, Malaluan R, Hirose S, Takeda N, Adschiri T, Arai K (1998) Cellulose hydrolysis in subcritical and supercritical water. J Supercrit Fluids 13(1):261–268

125. Sasaki M, Furukawa M, Minami K, Adschiri T, Arai K (2002) Kinetics and mechanism of cellobiose hydrolysis and retro-aldol condensation in subcritical and supercritical water. Ind Eng Chem Res 41(26):6642–6649

126. Harada T, Tokai Y, Kimura A, Ikeda S, Matsumura M (2014) Hydrolysis of crystalline cellulose to glucose in an autoclave containing both gaseous and liquid water. RSC Adv 4 (51):26838–26842

127. Rinaldi R, Palkovits R, Schüth F (2008) Depolymerization of cellulose using solid catalysts in ionic liquids. Angew Chem Int Ed 47(42):8047–8050

128. Rinaldi R, Meine N, vom Stein J, Palkovits R, Schüth F (2010) Which controls the depolymerization of cellulose in ionic liquids: the solid acid catalyst or cellulose? ChemSusChem 3(2):266–276

129. Zhang C, Fu Z, Liu YC, Dai B, Zou Y, Gong X, Wang Y, Deng X, Wu H, Xu Q, Steven KR, Yin D (2012) Ionic liquid-functionalized biochar sulfonic acid as a biomimetic catalyst for hydrolysis of cellulose and bamboo under microwave irradiation. Green Chem 14(7):1928–1934

130. Toda M, Takagaki A, Okamura M, Kondo JN, Hayashi S, Domen K, Hara M (2005) Green chemistry: biodiesel made with sugar catalyst. Nature 438(7065):178–178

131. Suganuma S, Nakajima K, Kitano M, Yamaguchi D, Kato H, Hayashi S, Hara M (2008) Hydrolysis of cellulose by amorphous carbon bearing SO₃H, COOH, and OH groups. J Am Chem Soc 130(38):12787–12793

132. Yamaguchi D, Kitano M, Suganuma S, Nakajima K, Kato H, Hara M (2009) Hydrolysis of cellulose by a solid acid catalyst under optimal reaction conditions. J Phys Chem C 113 (8):3181–3188

133. Fukuhara K, Nakajima K, Kitano M, Kato H, Hayashi S, Hara M (2011) Structure and catalysis of cellulose-derived amorphous carbon bearing SO₃H groups. ChemSusChem 4 (6):778–784

134. Mo X, López DE, Suwannakarn K, Liu Y, Lotero E, Goodwin JG Jr, Lu C (2008) Activation and deactivation characteristics of sulfonated carbon catalysts. J Catal 254(2):332–338

135. Kitano M, Yamaguchi D, Suganuma S, Nakajima K, Kato H, Hayashi S, Hara M (2009) Adsorption-enhanced hydrolysis of β-1,4-glucan on graphene-based amorphous carbon bearing SO₃H, COOH, and OH groups. Langmuir 25(9):5068–5075

136. Nakajima K, Hara M (2012) Amorphous carbon with SO₃H groups as a solid Brønsted acid catalyst. ACS Catal 2(7):1296–1304

137. Hara M, Yamaguchi D (2009) Method of hydrolyzing polysaccharide and stirring apparatus therefor. WO Patent 2009099218

138. Onda A, Ochi T, Yanagisawa K (2008) Selective hydrolysis of cellulose into glucose over solid acid catalysts. Green Chem 10(10):1033–1037

139. Onda A, Ochi T, Yanagisawa K (2009) Hydrolysis of cellulose selectively into glucose over sulfonated activated-carbon catalyst under hydrothermal conditions. Top Catal 52(6):801–807

140. Van de Vyver S, Peng L, Geboers J, Schepers H, de Clippel F, Gommes CJ, Goderis B, Jacobs PA, Sels BF (2010) Sulfonated silica/carbon nanocomposites as novel catalysts for hydrolysis of cellulose to glucose. Green Chem 12(9):1560–1563

141. Pang J, Wang A, Zheng M, Zhang T (2010) Hydrolysis of cellulose into glucose over carbons sulfonated at elevated temperatures. Chem Commun 46(37):6935–6937

142. Chung P-W, Charmot A, Gazit OM, Katz A (2012) Glucan adsorption on mesoporous carbon nanoparticles: effect of chain length and internal surface. Langmuir 28(43):15222–15232

143. Chung P-W, Charmot A, Click T, Lin Y, Bae YJ, Chu J-W, Katz A (2015) Importance of internal porosity for glucan adsorption in mesoporous carbon materials. Langmuir 31 (26):7288–7295

144. D-m Lai, Deng L, Li J, Liao B, Guo Q-x FuY (2011) Hydrolysis of cellulose into glucose by magnetic solid acid. ChemSusChem 4(1):55–58

145. D-m Lai, Deng L, Guo Q-x FuY (2011) Hydrolysis of biomass by magnetic solid acid. Energy Environ Sci 4(9):3552–3557

146. Takagaki A, Nishimura M, Nishimura S, Ebitani K (2011) Hydrolysis of sugars using magnetic silica nanoparticles with sulfonic acid groups. Chem Lett 40(10):1195–1197
147. Shuai L, Pan X (2012) Hydrolysis of cellulose by cellulase-mimetic solid catalyst. Energy Environ Sci 5(5):6889–6894
148. Merrifield RB (1963) Solid phase peptide synthesis. I. The synthesis of a tetrapeptide. J Am Chem Soc 85(14):2149–2154
149. Kobayashi H, Yabushita M, Hasegawa J, Fukuoka A (2015) Synergy of vicinal oxygenated groups of catalysts for hydrolysis of cellulosic molecules. J Phys Chem C 119(36):20993–20999
150. Kobayashi H, Komanoya T, Hara K, Fukuoka A (2010) Water-tolerant mesoporous-carbon-supported ruthenium catalysts for the hydrolysis of cellulose to glucose. ChemSusChem 3(4):440–443
151. Komanoya T, Kobayashi H, Hara K, Chun W-J, Fukuoka A (2011) Catalysis and characterization of carbon-supported ruthenium for cellulose hydrolysis. Appl Catal A Gen 407(1–2):188–194
152. Loerbroks C, Rinaldi R, Thiel W (2013) The electronic nature of the 1,4-β-glycosidic bond and its chemical environment: DFT insights into cellulose chemistry. Chem Eur J 19 (48):16282–16294
153. Zhao X, Wang J, Chen C, Huang Y, Wang A, Zhang T (2014) Graphene oxide for cellulose hydrolysis: how it works as a highly active catalyst? Chem Commun 50(26):3439–3442
154. Chung P-W, Charmot A, Olatunji-Ojo OA, Durkin KA, Katz A (2013) Hydrolysis catalysis of *Miscanthus* xylan to xylose using weak-acid surface sites. ACS Catal 4(1):302–310
155. To AT, Chung P-W, Katz A (2015) Weak-acid sites catalyze the hydrolysis of crystalline cellulose to glucose in water: importance of post-synthetic functionalization of the carbon surface. Angew Chem Int Ed 54(38):11050–11053
156. Gazit OM, Charmot A, Katz A (2011) Grafted cellulose strands on the surface of silica: effect of environment on reactivity. Chem Commun 47(1):376–378
157. Gazit OM, Katz A (2011) Grafted poly(1 → 4-β-glucan) strands on silica: a comparative study of surface reactivity as a function of grafting density. Langmuir 28(1):431–437
158. Gazit OM, Katz A (2013) Understanding the role of defect sites in glucan hydrolysis on surfaces. J Am Chem Soc 135(11):4398–4402
159. Balandin AA, Vasyunina NA, Barysheva GS, Chepigo SV, Dubinin MM (1957) Hydrogenation catalysts for polysaccharides. Bull Acad Sci USSR Div Chem Sci 6 (3):403–404
160. Balandin AA, Vasunina NA, Chepigo SV, Barysheva GS (1959) Hydrolytic hydrogenation of cellulose. Dokl Akad Nauk SSSR 128(5):941–944
161. Kobayashi H, Hosaka Y, Hara K, Feng B, Hirosaki Y, Fukuoka A (2014) Control of selectivity, activity and durability of simple supported nickel catalysts for hydrolytic hydrogenation of cellulose. Green Chem 16(2):637–644
162. Negahdar L, Oltmanns JU, Palkovits S, Palkovits R (2014) Kinetic investigation of the catalytic conversion of cellobiose to sorbitol. Appl Catal B Environ 147:677–683
163. Jacobs P, Hinnekens H (1990) Single-step catalytic process for the direct conversion of polysaccharides to polyhydric alcohols. US Patent 4,950,812
164. Deguchi S, Tsujii K, Horikoshi K (2006) Cooking cellulose in hot and compressed water. Chem Commun 31:3293–3295
165. Fukuoka A, Dhepe PL (2006) Catalytic conversion of cellulose into sugar alcohols. Angew Chem Int Ed 45(31):5161–5163
166. Fukuhara H, Matsunaga F, Yasuhara M, Araki S, Isaka T (1993) Preparation of propylene by dehydration of isopropanol in the presence of a pseudo-boehmite derived gamma alumina catalyst. US Patent 5,227,563
167. Jollet V, Chambon F, Rataboul F, Cabiac A, Pinel C, Guillon E, Essayem N (2009) Non-catalyzed and Pt/γ-Al$_2$O$_3$-catalyzed hydrothermal cellulose dissolution–conversion: influence of the reaction parameters and analysis of the unreacted cellulose. Green Chem 11 (12):2052–2060

168. Jollet V, Chambon F, Rataboul F, Cabiac A, Pinel C, Guillon E, Essayem N (2010) Non-catalyzed and Pt/γ-Al₂O₃ catalyzed hydrothermal cellulose dissolution-conversion: influence of the reaction parameters. Top Catal 53(15):1254–1257

169. Liu M, Deng W, Zhang Q, Wang Y, Wang Y (2011) Polyoxometalate-supported ruthenium nanoparticles as bifunctional heterogeneous catalysts for the conversions of cellobiose and cellulose into sorbitol under mild conditions. Chem Commun 47(34):9717–9719

170. Kobayashi H, Ito Y, Komanoya T, Hosaka Y, Dhepe PL, Kasai K, Hara K, Fukuoka A (2011) Synthesis of sugar alcohols by hydrolytic hydrogenation of cellulose over supported metal catalysts. Green Chem 13(2):326–333

171. Ravenelle RM, Diallo FZ, Crittenden JC, Sievers C (2012) Effects of metal precursors on the stability and observed reactivity of Pt/γ-Al₂O₃ catalysts in aqueous phase reactions. ChemCatChem 4(4):492–494

172. Yan N, Zhao C, Luo C, Dyson PJ, Liu H, Kou Y (2006) One-step conversion of cellobiose to C₆-alcohols using a ruthenium nanocluster catalyst. J Am Chem Soc 128(27):8714–8715

173. Luo C, Wang S, Liu H (2007) Cellulose conversion into polyols catalyzed by reversibly formed acids and supported ruthenium clusters in hot water. Angew Chem Int Ed 46 (40):7636–7639

174. Sweeton FH, Mesmer RE, Baes Jr CF (1974) Acidity measurements at elevated temperatures. VII. Dissociation of water. J Solution Chem 3 (3):191–214

175. Marshall WL, Franck EU (1981) Ion product of water substance, 0–1000 ℃, 1–10,000 bars New International Formulation and its background. J Phys Chem Ref Data 10(2):295–304

176. Geboers J, Van de Vyver S, Carpentier K, Jacobs P, Sels B (2011) Hydrolytic hydrogenation of cellulose with hydrotreated caesium salts of heteropoly acids and Ru/C. Green Chem 13 (8):2167–2174

177. Palkovits R, Tajvidi K, Ruppert AM, Procelewska J (2011) Heteropoly acids as efficient acid catalysts in the one-step conversion of cellulose to sugar alcohols. Chem Commun 47 (1):576–578

178. Ding L-N, Wang A-Q, Zheng M-Y, Zhang T (2010) Selective transformation of cellulose into sorbitol by using a bifunctional nickel phosphide catalyst. ChemSusChem 3(7):818–821

179. Yang P, Kobayashi H, Hara K, Fukuoka A (2012) Phase change of nickel phosphide catalysts in the conversion of cellulose into sorbitol. ChemSusChem 5(5):920–926

180. Van de Vyver S, Geboers J, Dusselier M, Schepers H, Vosch T, Zhang L, Van Tendeloo G, Jacobs PA, Sels BF (2010) Selective bifunctional catalytic conversion of cellulose over reshaped Ni particles at the tip of carbon nanofibers. ChemSusChem 3(6):698–701

181. Van de Vyver S, Geboers J, Schutyser W, Dusselier M, Eloy P, Dornez E, Seo JW, Courtin CM, Gaigneaux EM, Jacobs PA, Sels BF (2012) Tuning the acid/metal balance of carbon nanofiber-supported nickel catalysts for hydrolytic hydrogenation of cellulose. ChemSusChem 5(8):1549–1558

182. Pang J, Wang A, Zheng M, Zhang Y, Huang Y, Chen X, Zhang T (2012) Catalytic conversion of cellulose to hexitols with mesoporous carbon supported Ni-based bimetallic catalysts. Green Chem 14(3):614–617

183. Shrotri A, Tanksale A, Beltramini JN, Gurav H, Chilukuri SV (2012) Conversion of cellulose to polyols over promoted nickel catalysts. Catal Sci Technol 2(9):1852–1858

184. Kobayashi H, Matsuhashi H, Komanoya T, Hara K, Fukuoka A (2011) Transfer hydrogenation of cellulose to sugar alcohols over supported ruthenium catalysts. Chem Commun 47(8):2366–2368

185. Komanoya T, Kobayashi H, Hara K, Chun W-J, Fukuoka A (2014) Kinetic study of catalytic conversion of cellulose to sugar alcohols under low-pressure hydrogen. ChemCatChem 6 (1):230–236

186. Komanoya T, Kobayashi H, Hara K, Chun W-J, Fukuoka A (2013) Simultaneous formation of sorbitol and gluconic acid from cellobiose using carbon-supported ruthenium catalysts. J Energy Chem 22(2):290–295

187. Cook JD (2005) Diagnosis and management of iron-deficiency anaemia. Best Pract Res Clin Haematol 18(2):319–332

188. Stankiewicz BA, Briggs DEG, Evershed RP, Flannery MB, Wuttke M (1997) Preservation of chitin in 25-million-year-old fossils. Science 276(5318):1541–1543

189. Jang M-K, Kong B-G, Jeong Y-I, Lee CH, Nah J-W (2004) Physicochemical characterization of α-chitin, β-chitin, and γ-chitin separated from natural resources. J Polym Sci Part A Polym Chem 42(14):3423–3432

190. Khor E, Lim LY (2003) Implantable applications of chitin and chitosan. Biomaterials 24 (13):2339–2349

191. Bloch R, Burger MM (1974) Purification of wheat germ agglutinin using affinity chromatography on chitin. Biochem Biophys Res Commun 58(1):13–19

192. Krajewska B (2004) Application of chitin- and chitosan-based materials for enzyme immobilizations: a review. Enzyme Microbiol Technol 35(2):126–139

193. White RJ, Antonietti M, Titirici M-M (2009) Naturally inspired nitrogen doped porous carbon. J Mater Chem 19(45):8645–8650

194. Lavall RL, Assis OBG, Campana-Filho SP (2007) β-Chitin from the pens of *Loligo* sp.: extraction and characterization. Bioresour Technol 98(13):2465–2472

195. Nakagawa YS, Oyama Y, Kon N, Nikaido M, Tanno K, Kogawa J, Inomata S, Masui A, Yamamura A, Kawaguchi M, Matahira Y, Totani K (2011) Development of innovative technologies to decrease the environmental burdens associated with using chitin as a biomass resource: mechanochemical grinding and enzymatic degradation. Carbohydr Polym 83 (4):1843–1849

196. Osada M, Miura C, Nakagawa YS, Kaihara M, Nikaido M, Totani K (2012) Effect of sub- and supercritical water pretreatment on enzymatic degradation of chitin. Carbohydr Polym 88(1):308–312

197. Osada M, Miura C, Nakagawa YS, Kaihara M, Nikaido M, Totani K (2013) Effects of supercritical water and mechanochemical grinding treatments on physicochemical properties of chitin. Carbohydr Polym 92(2):1573–1578

198. Grigorian A, Araujo L, Naidu NN, Place DJ, Choudhury B, Demetriou M (2011) *N*-Acetylglucosamine inhibits T-helper 1 (Th1)/T-helper 17 (Th17) cell responses and treats experimental autoimmune encephalomyelitis. J Biol Chem 286(46):40133–40141

199. Salvatore S, Heuschkel R, Tomlin S, Davies S, Edwards S, Walker-Smith JA, French I, Murch SH (2000) A pilot study of N-acetyl glucosamine, a nutritional substrate for glycosaminoglycan synthesis, in paediatric chronic inflammatory bowel disease. Aliment Pharm Ther 14(12):1567–1579

200. Álvarez-Añorve LI, Calcagno ML, Plumbridge J (2005) Why does *Escherichia coli* grow more slowly on glucosamine than on *N*-acetylglucosamine? Effects of enzyme levels and allosteric activation of GlcN6P deaminase (NagB) on growth rates. J Bacteriol 187(9):2974–2982

201. Shikhman AR, Kuhn K, Alaaeddine N, Lotz M (2001) *N*-Acetylglucosamine prevents IL-1 β-mediated activation of human chondrocytes. J Immunol 166(8):5155–5160

202. Sashiwa H, Fujishima S, Yamano N, Kawasaki N, Nakayama A, Muraki E, Hiraga K, Oda K, Aiba S (2002) Production of *N*-acetyl-D-glucosamine from α-chitin by crude enzymes from *Aeromonas hydrophila* H-2330. Carbohydr Res 337(8):761–763

203. Einbu A, Vårum KM (2007) Depolymerization and de-N-acetylation of chitin oligomers in hydrochloric acid. Biomacromolecules 8(1):309–314

204. Hennen WJ, Sweers HM, Wang YF, Wong CH (1988) Enzymes in carbohydrate synthesis. Lipase-catalyzed selective acylation and deacylation of furanose and pyranose derivatives. J Org Chem 53(21):4939–4945

205. Omari KW, Dodot L, Kerton FM (2012) A simple one-pot dehydration process to convert *N*-acetyl-D-glucosamine into a nitrogen-containing compound, 3-acetamido-5-acetylfuran. ChemSusChem 5(9):1767–1772

206. Drover MW, Omari KW, Murphy JN, Kerton FM (2012) Formation of a renewable amide, 3-acetamido-5-acetylfuran, *via* direct conversion of N-acetyl-D-glucosamine. RSC Adv 2(11):4642–4644

207. Osada M, Kikuta K, Yoshida K, Totani K, Ogata M, Usui T (2013) Non-catalytic synthesis of Chromogen I and III from N-acetyl-D-glucosamine in high-temperature water. Green Chem 15(10):2960–2966

208. Osada M, Kikuta K, Yoshida K, Totani K, Ogata M, Usui T (2014) Non-catalytic dehydration of N,N'-diacetylchitobiose in high-temperature water. RSC Adv 4(64):33651–33657

209. Ohmi Y, Nishimura S, Ebitani K (2013) Synthesis of α-amino acids from glucosamine-HCl and its derivatives by aerobic oxidation in water catalyzed by Au nanoparticles on basic supports. ChemSusChem 6(12):2259–2262

210. Bobbink FD, Zhang J, Pierson Y, Chen X, Yan N (2015) Conversion of chitin derived N-acetyl-D-glucosamine (NAG) into polyols over transition metal catalysts and hydrogen in water. Green Chem 17(2):1024–1031

211. Chen X, Chew SL, Kerton FM, Yan N (2014) Direct conversion of chitin into a N-containing furan derivative. Green Chem 16(4):2204–2212

212. Pierson Y, Chen X, Bobbink FD, Zhang J, Yan N (2014) Acid-catalyzed chitin liquefaction in ethylene glycol. ACS Sustainable Chem Eng 2(8):2081–2089

213. Charmot A, Chung P-W, Katz A (2014) Catalytic hydrolysis of cellulose to glucose using weak-acid surface sites on postsynthetically modified carbon. ACS Sustainable Chem Eng 2(12):2866–2872

Chapter 2
Hydrolysis of Cellulose to Glucose Using Carbon Catalysts

2.1 Introduction

Cellulose is a potential and attractive alternative to petrol to reduce emission of greenhouse gases as this plant-derived biomass resource is an abundant, non-food, and renewable carbon source [1–6]. Glucose, a monomer of cellulose, is a key intermediate of various useful chemicals such as polymers, medicines, surfactants, gasoline, and diesel fuels (Sect. 1.2.3) [7–9]. Accordingly, the reaction of cellulose to glucose (Fig. 2.1) will be a mainstream in the next-generation biorefinery, which will substitute current processes using food biomass. However, realizing this vision has been hampered by the recalcitrance of cellulose and various practical difficulties as described in Chap. 1.

The hydrolysis of cellulose has been performed with various solid catalysts, as heterogeneous catalysts are advantageous over homogeneous ones in terms of easy separation from products [6, 10–12]. In the previous works, it was found that unmodified mesoporous carbon CMK-3 catalyzed the hydrolysis of cellulose [13, 14]. Fukuoka et al. has proposed that the active sites of CMK-3 would be weakly acidic groups [14]. Due to the absence of strong acid sites, CMK-3 stands in stark contrast to sulfonated carbons [15, 16] that have been widely studied in the hydrolysis of cellulose. Sulfonated catalysts possibly suffer from the leaching of SO_3H groups in hot compressed water [17], whereas weakly acidic carbons such as CMK-3 are expected to be highly hydrothermally stable. Furthermore, salts included in raw biomass easily deactivate strong acids by ion exchange; in contrast, weak acids potentially preserve their catalytic activities even in the presence of salts [18, 19]. These great advantages of weakly acidic carbons potentially break through the barrier of practical hydrolysis of cellulosic biomass. However, the highest glucose yield obtained by CMK-3 was only 16 %, and the synthesis of CMK-3 [20] was complicated in the practical perspective. Hence, in this chapter, the author has aimed the high-yielding synthesis of glucose from cellulose and real cellulosic biomass using more common carbon materials.

© Springer Science+Business Media Singapore 2016
M. Yabushita, *A Study on Catalytic Conversion of Non-Food Biomass into Chemicals*, Springer Theses, DOI 10.1007/978-981-10-0332-5_2

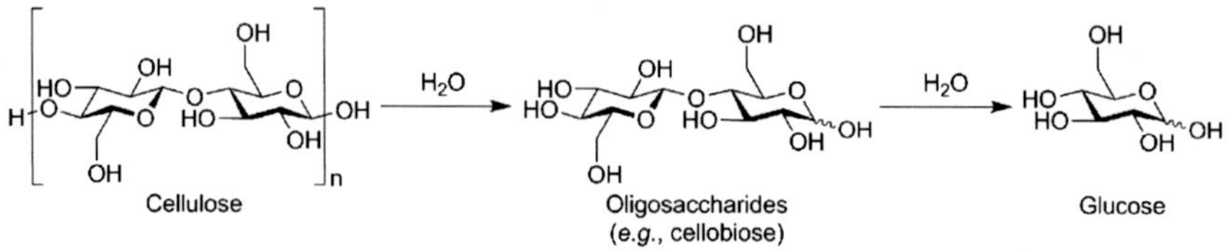

Fig. 2.1 Hydrolysis of cellulose to glucose via oligosaccharides

2.2 Experimental

2.2.1 Reagents

Microcrystalline cellulose	Column chromatography grade, 102331, Merck
Bagasse kraft pulp	Harvested in Okinawa Prefecture in Japan, composition determined by the established method (see Sect. 2.2.4) [21]: cellulose (59 wt%), hemicellulose [27 wt% (xylan 25 wt%, arabinan (2 wt%)], and lignin (9 wt%), Showa Denko
K26	Alkali-activated carbon (not for sale), Showa Denko
K20	Alkali-activated carbon (not for sale), Showa Denko
MSP20	Alkali-activated carbon, Kansai Coke & Chemicals
BA50	Steam-activated carbon, Ajinomoto Fine Techno
SX Ultra	Steam-activated carbon, Norit, denoted as SX
Vulcan XC72	Carbon black, Cabot, denoted as XC72
Black Pearls 2000	Carbon black, Cabot, denoted as BP2000
Pluronic P123	Triblock copolymer, $HO(CH_2CH_2O)_{20}(CH_2CH(CH_3)O)_{70}(CH_2CH_2O)_{20}H$, Sigma-Aldrich
Tetraethyl orthosilicate	>99.0 %, Sigma-Aldrich, denoted as TEOS
Ethanol	Special grade, Wako Pure Chemical Industries
Sulfuric acid	96–98 %, super special grade, Wako Pure Chemical Industries
Sucrose	Special grade, Wako Pure Chemical Industries
Hydrofluoric acid	46–48 %, special grade, Wako Pure Chemical Industries
Calcium gluconate	Special grade, Wako Pure Chemical Industries
Amberlyst 70	Sulfonic acid cation exchange resin, Organo
JRC-Z5-90H	Proton-type MFI zeolite, Si/Al ratio = 45, Catalysis Society of Japan, denoted as H-MFI
JRC-Z-HM90	Proton-type MOR zeolite, Si/Al ratio = 45, Catalysis Society of Japan, denoted as H-MOR

Q-6	Amorphous silica, Fuji Silysia Chemical, denoted as SiO_2
Silica-alumina	Grade 135, Sigma-Aldrich, denoted as SiO_2-Al_2O_3
JRC-TIO-4(2)	Titania, Catalysis Society of Japan, denoted as TiO_2
Hydrochloric acid	35–37 %, special grade, Wako Pure Chemical Industries
Acetic acid	>99.7 %, special grade, Wako Pure Chemical Industries
Sodium acetate	Special grade, Wako Pure Chemical Industries
Sodium chloride	Special grade, Wako Pure Chemical Industries
Benzoic acid	Special grade, Wako Pure Chemical Industries
Cellohexaose	>95 %, Seikagaku Biobusiness
Cellopentaose	>95 %, Seikagaku Biobusiness
Cellotetraose	>97 %, Seikagaku Biobusiness
Cellotriose	>97 %, Seikagaku Biobusiness
D(+)-Cellobiose	Special grade, Kanto Chemical
D(+)-Glucose	Special grade, Kanto Chemical
D(+)-Mannose	Special grade, Wako Pure Chemical Industries
D(−)-Fructose	Special grade, Kanto Chemical
1,6-Anhydro-β-D-glucopyranose	99 %, Wako Pure Chemical Industries, denoted as levoglucosan
5-Hydroxymethylfurfural	99 %, Sigma-Aldrich, denoted as 5-HMF
Xylobiose	Biochemical grade, Wako Pure Chemical Industries
D(+)-Xylose	Special grade, Wako Pure Chemical Industries
D(−)-Arabinose	Special grade, Wako Pure Chemical Industries
Sodium sulfide	Special grade, Wako Pure Chemical Industries
Distilled water	Wako Pure Chemical Industries
Distilled water	For HPLC, Wako Pure Chemical Industries
Acetonitrile	For HPLC, Wako Pure Chemical Industries
Milli-Q water	Prepared by an ultrapure water production system (Sartorius, arium 611UF)
Lithium chloride	Special grade, Wako Pure Chemical Industries
N,N-Dimethylacetamide	Special grade, Wako Pure Chemical Industries, denoted as DMAc
Methanol	Special grade, Wako Pure Chemical Industries
Deuterium oxide	For NMR, Acros Organics
Deuterated methanol	For NMR, Acros Organics
Calcium carbonate	Special grade, Wako Pure Chemical Industries
Sodium hydrogen carbonate	Special grade, Wako Pure Chemical Industries
Sodium carbonate solution	For volumetric analysis, 0.05 M, Wako Pure Chemical Industries
Sodium hydroxide solution	For volumetric analysis, 0.05 M, Wako Pure Chemical Industries

Hydrochloric acid solution   For volumetric analysis, 0.05 M, Wako Pure Chemical Industries
Methyl orange   Special grade, Wako Pure Chemical Industries
Nitrogen gas   Alpha gas 2, Air Liquide Kogyo Gas
Oxygen gas   Alpha gas 2, Air Liquide Kogyo Gas
Argon gas   Alpha gas 2, Air Liquide Kogyo Gas

2.2.2 Synthesis of Mesoporous Carbon CMK-3

Mesoporous carbon CMK-3 was synthesized by following the literature [20]. Pluronic P123 (12 g) was completely dissolved in an HCl aqueous solution (1.6 M, 450 g) at 308 K, and subsequently TEOS (26 g) was added dropwise into the solution over *ca.* 3 min under vigorous stirring using a magnetic stir bar. The agitation was further continued for 15 min at 308 K. The sample was aged at 308 K for 24 h and then at 373 K for 24 h in an oven without stirring. During stirring and aging, the lid of vessel was kept closed. The resulting solid and liquid were separated by suction filtration using a Buchner funnel with filter paper (4A, hard type) and a filtering flask with ethanol as an antiform. The solid was washed with distilled water (1.5 L) and was dried at 373 K through the night in an oven. The resulting powder was calcined in an electric furnace (Yamato, FO610) at 833 K for 8 h, for which the ramping rate of temperature was 10 K min^{-1}. After the calcination, 6.7 g of white powder (mesoporous silica SBA-15) was obtained.

Synthesized SBA-15 (2.0 g) was dispersed in a sucrose aqueous solution [sucrose (2.5 g), H$_2$SO$_4$ (0.29 g), and distilled water (10 mL)], and the suspension was stirred at room temperature for 1 h to impregnate sucrose into the pores of SBA-15. After the filtration, the powder was dried at 373 K for 6 h and then at 433 K for 6 h. The powder was crashed on a mortar and was again dispersed in a sucrose solution [sucrose (1.6 g), H$_2$SO$_4$ (0.19 g), and distilled water (10 mL)]. After filtration and drying under the same conditions described above, the powder was heated to 1173 K by 2.4 K min^{-1}, and then the temperature was kept for 6 h in a horizontal quartz tube [inner diameter ø30 mm, temperature was controlled by an electric furnace (AZ ONE, TMF-300 N)] under N$_2$ flow (10 mL min^{-1}) to carbonize impregnated sucrose. The carbon/SBA-15 composite was dispersed in an HF aqueous solution (10 wt%, 100 g), and the suspension was stirred at room temperature for 4 h to dissolve the SBA-15 template. The suspension was separated by filtration using a Buchner funnel with filter paper (4A, hard type) and a filtering flask containing a calcium gluconate solution to quench HF. The resulting powder was washed with distilled water repeatedly and dried in an oven at 373 K overnight. Then, 1.2 g of black powder (mesoporous carbon CMK-3) was obtained.

2.2.3 Quantitative Analysis of OFGs on Carbon Materials

Carboxylic acids, lactones, and phenolic groups on carbon materials were quantified by the Boehm titration [22]. 0.5 g of carbon was dispersed in a base aqueous solution (50 mM, 20 mL, containing $NaHCO_3$, Na_2CO_3, or $NaOH$). The suspension was stirred at 600 rpm at 298 K for 24 h under Ar, and the liquid and solid were separated by filter paper (5A, quantitative) to remove carbon. Twice of a stoichiometric amount of HCl aqueous solution (50 mM) against base used was added to 5 mL of the filtrate. After sonication under 500 hPa of reduced pressure for 5 min to remove CO_2 dissolving in the solution, back titration was conducted using a 50 mM of NaOH aqueous solution with methyl orange as an indicator. Note that phenolphthalein is not suitable as an indicator since equilibrium of carbonate ion influences color change of phenolphthalein at high pH.

2.2.4 Analysis of Components in Bagasse Kraft Pulp

The weight ratio of cellulose, hemicellulose (xylan and arabinan), and lignin in bagasse kraft pulp was determined by following the established technique, so-called National Renewable Energy Laboratory (NREL) method [21]. 300 mg of bagasse kraft pulp was dispersed in 3 mL of H_2SO_4 aqueous solution (72 wt%), and the suspension was stirred at 303 K for 1 h. After adding 84 mL of distilled water to dilute H_2SO_4, the mixture was reacted at 394 K for 1 h in a hastelloy C22 high-pressure reactor (OM Lab-Tech, MMJ-100, 100 mL, Fig. 2.2). The solid and liquid phases were separated by filtration, and then $CaCO_3$ was added to the liquid phase to quench H_2SO_4 to be pH 7. The pH should be kept ≤7 to prevent sugars from their decomposition. The resulting solution was analyzed by high-performance liquid chromatography (HPLC, the conditions are shown in Sect. 2.2.6) to determine the amounts of glucose, xylose, and arabinose, corresponding to cellulose, xylan, and arabinan, respectively. The solid was washed with distilled water and dried in an oven at 378 K overnight and then was calcined in an electric furnace (Denken-Highdental, KDF-S90) at 848 K for 24 h under air. The amount of lignin was determined from weight decrease by the calcination.

2.2.5 Ball-Milling and Mix-Milling Pretreatment of Cellulose

Microcrystalline cellulose (10 g) was ball-milled in the presence of ZrO_2 balls (ø1.0 cm, 1 kg) in a ceramic pot (0.9 L) at 60 rpm for 96 h or in the presence of Al_2O_3 balls (ø1.5 cm, 2 kg) in a ceramic pot (3.6 L) at 60 rpm for 48 h. Mix-milling

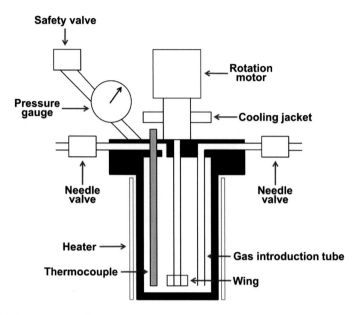

Fig. 2.2 Diagram of hastelloy C22 high-pressure reactor

of cellulose and solid catalyst was carried out in the same type of pot with the Al_2O_3 balls. Microcrystalline cellulose (10 g) and solid catalyst (1.54 g) (S/C ratio based on weight = 6.5) were added into the pot and were milled together at 60 rpm for 48 h. The amount of catalyst was reduced to 1.46 g for the mix-milling of cellobiose (S/C = 6.8).

Ball-milled samples were analyzed by XRD (Rigaku, MiniFlex, Cu Kα radiation), ^{13}C cross polarization/magic angle spinning NMR (^{13}C CP/MAS NMR, Bruker, MSL-300, 75 MHz, MAS frequency 8 kHz), laser diffraction (Nikkiso, Microtrac MT3300EXII, for estimation of secondary particle size), optical microscope (KEYENCE, VHX-5000), and scanning electron microscope (SEM, JEOL, JSM-6360LA). The viscometry was conducted at 303 K using an Ubbelohde viscometer (No. 0C) in 9 wt% LiCl/DMAc solvent [23]. The dissolution of cellulose in 9 wt% LiCl/DMAc was conducted by following the reported procedure [24]. 200 mg of cellulose sample was dispersed in 100 mL of Milli-Q water, and the suspension was stirred at room temperature overnight. The liquid phase was removed by centrifugation and decantation. The remaining solid was dispersed in 100 mL of methanol, and the suspension was stirred for 30 min at room temperature. After removing solvent by centrifugation and decantation, the solid was immersed in 50 mL of DMAc for 30 min at room temperature, and this procedure was repeated three times. Then, the resulting solid was dissolved in 9 wt% LiCl/DMAc. The prepared cellulose solution was used for the viscometry.

2.2.6   Catalytic Hydrolysis of Cellulosic Molecules

The hydrolysis of cellulose was conducted in the hastelloy C22 high-pressure reactor (Fig. 2.2). Ball-milled cellulose (324 mg), catalyst (50 mg), and distilled water (40 mL) were charged into the reactor. For the hydrolysis of mix-milled samples, 374 mg of the sample [containing cellulose (324 mg) and catalyst (50 mg)] and distilled water (40 mL) were used. The reactor was heated to 503 K in 18 min and then cooled to 323 K by blowing air for 22 min (named rapid heating-cooling condition, Fig. 2.3). The suspension was separated by centrifugation and decantation. The products in the aqueous phase were analyzed by HPLC [Shimadzu, LC10-ATVP, equipped with refractive index (RI) and ultraviolet (UV, 210 nm) detectors as well as a fraction collector] with a SUGAR SH1011 column (Shodex, ø8 × 300 mm, mobile phase: water at 0.5 mL min^{-1}, 323 K), a Rezex RPM-Monosaccharide Pb++ column (Phenomenex, ø7.8 × 300 mm, mobile phase: water at 0.6 mL min^{-1}, 343 K), and a TSKgel Amide-80 HR column [Tosoh, ø4.6 × 250 mm, mobile phase: acetonitrile/water (6/4, vol/vol) at 0.8 mL min^{-1}, 303 K]. An absolute calibration method was employed for calculation of product yields (Eq. 2.1). The conversion of cellulose was determined based on the weight difference of the solid part before and after reaction (Eq. 2.2).

$$\left(\text{Yield}/\%\right) = \frac{CF_{\text{HPLC}} \times A_{\text{HPLC}} \times V}{\frac{M_{\text{cellulose}}}{162.13}} \times \frac{n}{6} \times 100 \tag{2.1}$$

where CF_{HPLC} (unit: M count^{-1}) is calibration factor, A_{HPLC} (count) is peak area in HPLC chart, V (L) is volume of reaction solution, $M_{\text{cellulose}}$ (g) is mass of cellulose charged, and n is carbon number in a product.

$$\left(\text{Conversion}/\%\right) = \frac{M_{\text{residue}} - M_{\text{catalyst}}}{M_{\text{cellulose}}} \times 100 \tag{2.2}$$

Fig. 2.3 Temperature profile of rapid heating-cooling condition

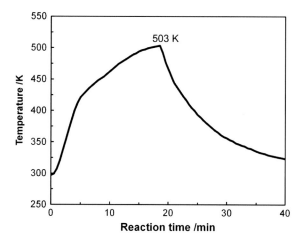

where $M_{residue}$ (unit: g) is mass of dried residue and $M_{catalyst}$ (g) is mass of catalyst charged.

The amount of organic carbons in reaction solution was quantified by measurement of total organic carbon (TOC, Shimadzu TOC-V_{CSN}) to determine conversion (Eq. 2.3) when catalyst was partially dissolved in water after the reaction (see Sect. 2.3.2).

$$(Conversion/\%) = \frac{CF_{TOC} \times A_{TOC}}{M_{cellulose} \times \frac{72.06}{162.13}} \times 100 \qquad (2.3)$$

where CF_{TOC} (unit: gram-carbon count^{-1}) is calibration factor and A_{TOC} (count) is peak area in TOC measurement.

The products were identified by HPLC, liquid chromatography/mass spectroscopy [LC/MS, Thermo Fischer Scientific, LCQ Fleet, atmospheric pressure chemical ionization (APCI), the conditions were the same as those of HPLC], and ^{1}H NMR spectroscopy (JEOL, JNM-ECX400, 400 MHz).

The hydrolysis of mix-milled samples at a lower temperature ($\leq$423 K) was carried out in a pressure-resistant glass tube (Ace Glass, 15 mL). 94 mg of mix-milled sample [containing cellulose (81 mg) and catalyst (13 mg)] and distilled water (10 mL) were charged into the tube. The tube was immersed in an oil bath at a certain temperature for a designated length of time. The product analysis was performed using the same procedure described above.

The hydrolysis of cellobiose was conducted in the hastelloy C22 high-pressure reactor. Cellobiose (342 mg), catalyst (50 mg), and distilled water (40 mL) were charged into the reactor. The temperature was raised to 463 K in 11 min, and then the reactor was rapidly cooled to 323 K by blowing air. The product yield and conversion were determined by HPLC (vide supra).

2.2.7   Reuse Test of Mix-Milled Cellulose

Microcrystalline cellulose (10 g) and K26 (1.54 g) were ball-milled together in the presence of the Al_2O_3 balls (ø1.5 cm, 2 kg) in the ceramic pot (3.6 L) at 60 rpm for 48 h. The recovered sample and 0.012 wt% HCl aqueous solution (40 mL) were charged into the hastelloy C22 high-pressure reactor (Fig. 2.2). The temperature was first raised to 473 K and was maintained for 2 min. Then, the reactor was cooled down to 423 K by blowing air and the mixture was further reacted for 60 min at the temperature. After cooling down to room temperature, the solid and liquid phases were separated by centrifugation and decantation. The solid phase was washed with distilled water repeatedly to extract physisorbed products, for which total volume of water used was *ca.* 160 mL. The resulting solid was dried under vacuum overnight at 353 K. The liquid phase and washing solvent were mixed and then analyzed by HPLC (vide supra). In the next run, the dried solid was mix-milled with fresh cellulose, the amount of which was the same as that of converted cellulose in the previous run. The experimental schemes are drawn in Sect. 2.3.2.

2.3 Results and Discussion

2.3.1 Screening of Carbon Catalysts for Hydrolysis of Cellulose

First, catalytic activity of various carbon materials has been investigated for the hydrolysis of individually ball-milled cellulose in water under the rapid heating–cooling condition at 503 K (Fig. 2.3), and Table 2.1 summarizes the reaction results. Among the carbon materials tested, alkali-activated carbon K26 was the most active catalyst and converted 60 % of cellulose (entry 2). The main product was glucose with 36 % yield, which was clearly higher than that in a control experiment without catalysts (4.6 %, entry 1). The other identified products were water-soluble oligosaccharides (2.5 %, DP = mainly 2–6), fructose (2.7 %), mannose (2.6 %), levoglucosan (2.1 %), and 5-HMF (3.4 %). Unidentified water-soluble compounds (16 %) were also formed, the yield of which was calculated from carbon balance. In repeated reactions, K26 maintained its catalytic performance at least four times.

Table 2.1 Hydrolysis of individually ball-milled cellulose by carbon catalysts

Entry	Catalyst	Conv./%	Yield based on carbon/%						
			Glucan		By-product				
			Glc[a]	Olg[b]	Frc[c]	Man[d]	Lev[e]	HMF[f]	Others[g]
1	None	28	4.6	15	0.5	0.6	0.2	1.8	5.3
2	K26	60	36	2.5	2.7	2.6	2.1	3.4	11
3	K20	59	35	1.7	1.9	1.3	2.7	2.9	14
4	MSP20	50	26	6.3	1.7	1.6	1.6	2.1	11
5	BA50	57	17	20	1.0	1.0	0.9	3.5	14
6	CMK-3	52	12	25	0.9	0.8	0.7	2.5	10
7	SX	38	8.1	20	0.6	0.9	0.4	1.4	6.6
8	BP2000	37	6.4	12	0.5	0.5	0.3	1.8	16
9	XC72	35	5.8	19	0.6	0.9	0.2	1.9	6.6

Conditions individually ball-milled cellulose (0.9 L pot) 324 mg; catalyst (not milled) 50 mg; distilled water 40 mL; 503 K; rapid heating-cooling condition (Fig. 2.3)
[a]Glucose
[b]Water-soluble oligosaccharides (DP = mainly 2–6)
[c]Fructose
[d]Mannose
[e]Levoglucosan
[f]5-HMF
[g](Conversion) – (Total yield of identified products)

Alkali-activated carbon K20 showed similar catalytic activity to K26, and gave 59 % conversion of cellulose and 35 % yield of glucose (entry 3). Another alkali-activated carbon MSP20, steam-activated carbon BA50, and mesoporous carbon CMK-3 were less active than K26 and K20 (entries 4–6). The other carbons, steam-activated carbon SX, carbon black BP2000, and carbon black XC72 were almost inactive (entries 7–9). These results show that catalytic activity of carbon materials strongly depends on their chemical and/or physical properties, derived from precursors and synthetic methods. Consequently, K26, which can be synthesized in a bulk scale in industry, showed a significantly higher catalytic activity than CMK-3, and thus K26 was mainly used for further study.

Next, the amounts of carboxylic acids, lactones, and phenolic groups as OFGs on carbons were quantified by the Boehm titration [22] (Table 2.2). The highly active catalysts such as K26 and K20 contain a large amount of OFGs (entries 10

Table 2.2 Amount of OFGs on carbon materials

Entry	Carbon	OFGs/µmol g^{-1}				Conv.^{a}/%	Glc^{a,b}/%
		Carboxylic acids	Lactones	Phenolic groups	Total		
10	K26	270	310	310	880	60	36
11	K20	420	270	320	1010	59	35
12	MSP20	65	110	280	450	50	26
13	CMK-3	140	97	180	410	52	12
14	SX	80	49	34	160	38	8.1
15	BP2000	61	12	88	160	37	6.4
16	XC72	26	49	0	75	35	5.8

Determined by the Boehm titration [22]
^{a}Reaction conditions: individually ball-milled cellulose (0.9 L pot) 324 mg; catalyst (not milled) 50 mg; distilled water 40 mL; 503 K; rapid heating-cooling condition (Fig. 2.3). The detailed results are summarized in Table 2.1
^{b}Glucose yield

Fig. 2.4 Effect of OFGs amount on catalytic activity of carbons

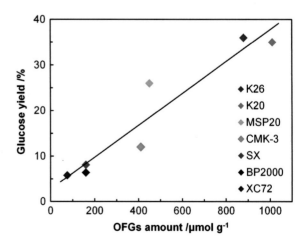

and 11), and less active catalysts do lower amounts (entries 12–16). The total amount of OFGs on carbons positively correlates with catalytic activity for glucose production from cellulose (Fig. 2.4), indicating that weakly acidic OFGs may be active sites. However, this trend would also include the influence of physical properties of each carbon material, and the detailed functions of OFGs are discussed in Chap. 3. It is notable that K26 does not contain sulfonic groups since the sulfur content is less than 0.01 %. This amount of sulfur corresponds to only $<10^{-5}$ % in the whole reaction mixture and does not catalyze the reaction at all. In fact, the hydrolysis in the presence of even 10^{-4} % H_2SO_4 provided 4.3 % yield of glucose (Table 2.3, entry 17) as low as the yield of blank experiment (4.6 %, entry 1). The author further conducted control experiments to evaluate acidity of active sites. K26 was treated in AcOH/AcONa (pH 4.0), NaCl (pH 7.0), NaHCO$_3$ (pH 8.3), and NaOH (pH 12.6) to neutralize corresponding acidic sites on K26 and the treated carbons were subjected to the hydrolysis of individually ball-milled cellulose (Fig. 2.5). The acetate buffer treatment did not influence the catalytic activity of K26, indicating that no strong acid worked as a predominant active site. Similarly, K26 has resisted an actual salt NaCl; K26 possibly maintains its catalytic activity even in the presence of salts. The NaHCO$_3$-treated K26 still produced glucose in 13 % yield, but exposure to NaOH completely deactivated K26 due to neutralization

Table 2.3 Control experiments on hydrolysis of individually ball-milled cellulose

Entry	Solvent	Conv./%	Yield based on carbon/%						
			Glucan		By-product				
			Glc[a]	Olg[b]	Frc[c]	Man[d]	Lev[e]	HMF[f]	Others[g]
1	Distilled water	28	4.6	15	0.5	0.6	0.2	1.8	5.3
17	10 μM H$_2$SO$_4$[h]	29	4.3	17	0.8	0.8	0.2	1.9	3.9
18	50 μM AcOH[i]	26	3.8	17	0.8	0.8	0.2	2.0	1.9
19	Filtrate of used K26[j]	22	3.1	13	0.5	0.6	0.1	1.4	2.9
20	Filtrate of used K26 and cellulose[k]	39	n.d.[l]	n.d.[l]	n.d.[l]	n.d.[l]	n.d.[l]	n.d.[l]	n.d.[l]

Conditions individually ball-milled cellulose (0.9 L pot) 324 mg; solvent *ca.* 40 mL; 503 K; rapid heating-cooling condition (Fig. 2.3)
[a]Glucose
[b]Water-soluble oligosaccharides (DP = mainly 2–6)
[c]Fructose
[d]Mannose
[e]Levoglucosan
[f]5-HMF
[g](Conversion) – (Total yield of identified products)
[h]pH = 4.7
[i]pH = 4.6
[j]The filtrate of K26 aqueous mixture subjected to the rapid heating–cooling condition (Fig. 2.3) was used as a solvent instead of distilled water
[k]The filtrate of the mixture of the cellulose hydrolysis by K26 (Table 2.1, entry 2) was used as a solvent instead of distilled water
[l]The values were unable to be determined because products formed from cellulose in the primary reaction (Table 2.1, entry 2) were contained, and they remained as themselves and/or underwent the degradation during this reaction

Fig. 2.5 Hydrolysis of individually ball-milled cellulose by base-treated K26. Conditions: individually ball-milled cellulose (0.9 L pot) 324 mg; catalyst 50 mg; distilled water 40 mL; 503 K; rapid heating-cooling condition (see Fig. 2.3)

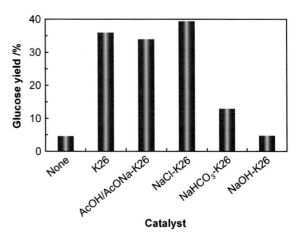

of all acid sites. Hence, the hydrolytic activity of K26 may be ascribed to weakly acidic groups such as carboxylic acids and phenolic groups. These results also suggest that weak acids are advantageous over strong acids owing to the resistance to salts derived from raw biomass since strong acids easily undergo ion exchange to be deactivated even in an acetate buffer [18, 19].

The author investigated whether the most active catalyst K26 worked as a *solid* catalyst in the hydrolysis of cellulose. The dispersion of K26 into distilled water caused the drop of pH to 4.9 due to weakly acidic OFGs. The same phenomenon is also observed when dispersing solid acids such as zeolites in water [25–27]. After filtration with a polytetrafluoroethylene (PTFE) membrane to remove K26, the pH value returned to almost neutral (5.8). This result confirmed that the pH decrease was not caused by leaching of soluble acidic species from K26. The acidic solutions with the same pH value as that of the K26 suspension, i.e., 10 μM H_2SO_4 (pH 4.7) and 50 μM AcOH (pH 4.6), were not active for the hydrolysis of cellulose at all (Table 2.3, entries 17 and 18). As it is known that the hydrolysis by H_3O^+ is negligible at pH higher than 4 [28], the promotion of the hydrolysis by K26 is not ascribed to the buffering effect releasing H_3O^+ in the suspended state. In addition to these evidences, the filtrates of the K26 suspension and K26-cellulose one, which were subjected to the reaction conditions at 503 K (Fig. 2.3), were less active (entries 19 and 20) than *solid* K26 itself (entry 2). Although the filtrate in entry 20 possibly contained soluble acidic by-products formed in the primary reaction of cellulose, the catalytic effects of the acidic species were limited in the hydrolysis of cellulose; they enhanced only 11 % of cellulose conversion from the blank reaction (entry 1). Hence, K26 should hydrolyze cellulose as a *solid* catalyst. The author also confirmed whether cellulose underwent hydrolysis as a *solid* substrate. The DP of ball-milled cellulose was 640, determined by viscometry (see Sect. 2.3.3), the value of which was high enough to be insoluble in water. The ball-milled cellulose did not contain a considerable amount of soluble oligomers (only 0.3 %), verified by the extraction with 40 mL of boiling water. Meanwhile, it has been reported that cellulose slightly dissolves in water at ambient temperature (*ca.* 2×10^{-3} wt%) [29] and complete dissolution of cellulose in water requires high temperatures as well as extremely high pressures (e.g., 603 K, 345 MPa) [30]. These reports motivated the author to estimate the influence of cellulose solubility on the hydrolysis at 503 K. If the slightly soluble portion of cellulose was responsible for the reaction, the hydrolysis rate should depend on the saturated solubility of cellulose, giving a constant concentration of dissolved cellulose. In contrast to the hypothesis, the conversion rate increased linearly with increasing amount of solid cellulose (Table 2.4 and Fig. 2.6). Although the possibility of soluble active species or partial dissolution of the substrate is not completely excluded, these results show that *solid* cellulose undergoes the hydrolysis catalyzed mainly by *solid* K26. Hence, both carbon and cellulose behave as solids in the hydrolysis, namely solid–solid reaction.

Table 2.4 Effect of cellulose amount on hydrolysis by K26

Entry	Amount of ball-milled cellulose/g L^{-1}	Conv./%	Conv. rate[a]/g L^{-1} h^{-1}	Yield based on carbon/%						
				Glucan		By-product				
				Glc[b]	Olg[c]	Frc[d]	Man[e]	Lev[f]	HMF[g]	Others[h]
21	8.1	72	8.7	40	1.4	2.1	1.8	2.8	5.4	19
22	16	66	16	40	1.5	1.7	1.3	2.5	6.3	13
23	24	66	24	39	1.7	1.4	1.2	2.6	7.0	13
24	40	58	35	38	2.1	1.3	1.2	2.2	6.7	7.2

Conditions K26 50 mg; distilled water 40 mL; 508 K; rapid heating-cooling condition (Fig. 2.3). Cellulose was ball milled without catalyst in a 0.9 L pot
[a]Conversion rate of cellulose based on the total reaction time (40 min, see Fig. 2.3)
[b]Glucose
[c]Water-soluble oligosaccharides (DP = mainly 2–6)
[d]Fructose
[e]Mannose
[f]Levoglucosan
[g]5-HMF
[h](Conversion) – (Total yield of identified products)

Fig. 2.6 Effect of cellulose amount on conversion rate

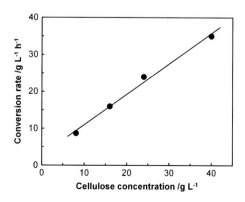

In solid–solid reactions, a collision between a solid substrate and a solid catalyst is limited, hampering reaction proceeding. In other words, solid catalysts would unsatisfactorily show their catalytic performance during reactions because of loose contact with a substrate.

2.3.2 High-Yielding Production of Glucose from Cellulose

In the previous section, K26 has provided glucose in 36 % yield from individually ball-milled cellulose (Table 2.1, entry 2), but the yield needs to be improved for practical use. The hydrolysis of cellulose by carbon catalyst takes place at their interface, namely solid–solid reaction; loose contact between a solid substrate and a solid catalyst is an obstacle. The author therefore employed a new pretreatment method named *mix-milling*, ball-milling cellulose and catalyst together, to make tight contact (Fig. 2.7).

The various solid materials including carbons and typical solid acids were tested for the hydrolysis of cellulose at 453 K for 20 min after the mix-milling pretreatment. In control experiments employing individually ball-milled cellulose as a substrate, the hydrolysis in the presence/absence of solid catalysts provided poor reaction results (7.9–15 % yields of glucans, Table 2.5, entries 25–34). The results of the hydrolysis of mix-milled cellulose are also shown in Table 2.5 (entries 35–43). K26 produced water-soluble glucans in 90 % yield [glucose (20 %) and oligosaccharides (70 %)] with 97 % selectivity (entry 35). The other products were fructose (0.6 %), mannose (0.7 %), levoglucosan (0.7 %), 5-HMF (1.0 %), and unidentified compounds (<1 %). The solid residue containing K26 was easily separated from the product cocktail by filtration after the reaction (Fig. 2.8). The mix-milling pretreatment also enhanced the yields of glucans when employing the other carbon materials, BA50 (34 %, entry 36) and SX (22 %, entry 37), compared to the individual milling (entries 27 and 28). However, these carbons were less active than K26 due to lower amounts of weakly acidic groups as described in Sect. 2.3.1. A sulfonic acid cation exchange resin Amberlyst 70 afforded remarkably high glucose yield (82 %, entry 38). After mix-milling of cellulose with Amberlyst 70, >99 wt% of the sample dissolved as various kinds of oligosaccharides [31]

Fig. 2.7 Schematic of mix-milling pretreatment

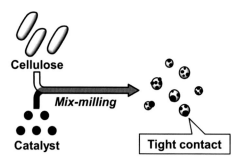

Cellulose

Mix-milling

Catalyst

Tight contact

Table 2.5   Effect of pretreatment on hydrolysis of cellulose by various solid catalysts

Entry	Catalyst	Pretreatment	Conv./%	Yield based on carbon/%						
				Glucan		By-product				
				Glc^a	Olg^b	Frc^c	Man^d	Lev^e	HMF^f	$Others^g$
25	None	Individual[k]	12	1.3	6.6	0.2	0.2	<0.1	0.2	3.4
26	K26	Individual[l]	18	2.9	10	0.5	0.4	0.1	<0.1	3.7
27	BA50	Individual[k]	20	2.4	8.6	0.2	0.3	<0.1	<0.1	8.0
28	SX	Individual[k]	16	2.3	8.0	0.3	0.3	<0.1	<0.1	4.8
29	Amberlyst 70	Individual[k]	18	6.3	8.2	0.2	0.8	0.2	0.3	2.4
30	H-MFI	Individual[k]	16	3.2	9.1	0.3	0.2	0.1	0.1	3.3
31	H-MOR	Individual[k]	17	3.8	9.6	0.3	0.3	0.1	0.2	3.0
32	$SiO_2\text{-}Al_2O_3$	Individual[k]	7.0	0.9	5.4	0.2	0.2	<0.1	0.2	0.1
33	SiO_2	Individual[k]	12	2.1	8.3	0.3	0.4	<0.1	0.2	1.1
34	TiO_2	Individual[k]	14	2.3	9.7	0.4	0.4	0.1	0.3	0.2
35	K26	Mix[m]	93	20	70	0.6	0.7	0.7	1.0	<1[n]
36	BA50	Mix[m]	35	6.7	27	0.7	0.7	0.2	0.2	0.2
37	SX	Mix[m]	24	4.2	18	0.8	0.5	0.1	0.3	0.1
38	Amberlyst 70	Mix[m]	>99	82	1.9	0.5	1.4	2.6	2.8	8.8
39	H-MFI	Mix[m]	19	4.0	11	0.4	0.3	0.1	0.2	3.2
40	H-MOR	Mix[m]	21	4.9	11	0.4	0.3	0.2	0.5	4.0
41	$SiO_2\text{-}Al_2O_3$	Mix[m]	6.8	0.9	4.8	0.2	0.2	<0.1	0.2	0.5
42	SiO_2	Mix[m]	16	3.4	11	0.3	0.3	0.1	0.3	0.1
43	TiO_2	Mix[m]	13	1.6	7.1	0.5	0.3	<0.1	0.4	2.6
44	K26, HCl[i]	Mix[m]	98	88	2.7	1.5	1.5	3.0	1.7	<1[n]
45	K26, $H_2SO_4^j$	Mix[m]	95	69	8.6	0.7	1.3	2.2	1.8	11
46	HCl[i]	Individual[j]	39	27	3.9	1.8	1.8	1.0	1.6	1.8
47	K26, HCl[i]	Individual[l]	40	30	4.3	1.3	1.3	1.1	0.7	0.7
48[h]	K26	Mix[m]	97	72	2.8	1.4	1.5	1.4	4.9	13

Conditions cellulose 324 mg; catalyst 50 mg; distilled water 40 mL; 453 K; 20 min. For hydrolysis of mix-milled cellulose, 374 mg of sample (containing cellulose 324 mg and catalyst 50 mg) was used
[a]Glucose
[b]Water-soluble oligosaccharides (DP = mainly 2–6)
[c]Fructose
[d]Mannose
[e]Levoglucosan
[f]5-HMF
[g](Conversion) – (Total yield of identified products)
[h]Conditions: mix-milled cellulose 94 mg (containing cellulose 81 mg and K26 13 mg); distilled water 10 mL; 418 K; 24 h
[i]Hydrolysis was conducted in a 0.012 wt% HCl solution (pH 2.5)
[j]Hydrolysis was conducted in a 0.018 wt% H_2SO_4 solution (pH 2.5)
[k]Cellulose was ball-milled without catalyst in a 3.6 L pot. Catalyst was not ball-milled
[l]Cellulose and K26 were separately ball-milled
[m]Cellulose and catalyst were ball-milled together, namely mix-milling
[n]The total yield of others was not correctly determined because of high conversion

Fig. 2.8 Pictures of **a** product cocktail and **b** solid residue after hydrolysis of mix-milled cellulose containing K26. The solid and liquid phases were separated with a PTFE membrane (0.1 μm mesh)

in 40 mL of distilled water at room temperature, indicating that mechanocatalytic hydrolysis of cellulose [31–33] happened during the milling process in the presence of strong acidic species derived from the resin. The water-soluble oligosaccharides are hydrolyzed more easily than robust cellulose, resulting in the high-yielding production of glucose. However, the reaction solution was completely homogeneous (Fig. 2.9), and Amberlyst 70 was unable to be recovered by centrifugation at

Laser

Fig. 2.9 Reaction mixture after hydrolysis of mix-milled cellulose containing Amberlyst 70. No solids were obtained after the reaction. The solution was irradiated with a green laser from right side of the bottle, but the Tyndall effect was hardly observed, showing that the solution was homogeneous and did not contain colloids

$4600g$ (4.5×10^4 m s^{-2}) or filtration using a PTFE membrane (0.1 μm mesh), which was in sharp contrast to the behavior of K26 is shown in Fig. 2.8. The other materials (H-MFI, H-MOR, SiO$_2$-Al$_2$O$_3$, SiO$_2$, and TiO$_2$) were almost inactive even though employing the mix-milling pretreatment (entries 39–43). Moreover, these materials except TiO$_2$ partially dissolved in water during the reaction (Fig. 2.10).

K26 was the most active catalyst in the hydrolysis of mix-milled cellulose among the solid materials tested except for Amberlyst 70. There are two possibilities for this: crystalline structure of cellulose contained in each mix-milled sample (see Sect. 1.2.2) as well as activity of each catalyst. To evaluate the influence of crystallinity of cellulose on the hydrolysis, all the mix-milled samples were characterized by XRD. The XRD patterns of all mix-milled samples (Fig. 2.11) were the same as that of amorphous cellulose (shown as individually milled cellulose in figure), and no peak derived from crystalline cellulose was observed. The mix-milling pretreatment similarly degraded crystalline cellulose to amorphous one regardless of the presence of various solids. Hence, the difference in reaction results is not ascribed to the nature of cellulose but to activity of each catalyst.

The mix-milled cellulose containing K26 gave glucose and water-soluble oligosaccharides in 20 and 70 % yields, respectively, as already shown above (Table 2.5, entry 35). In order to further improve glucose yield by accelerating depolymerization of oligosaccharides to glucose, a 0.012 % HCl aqueous solution

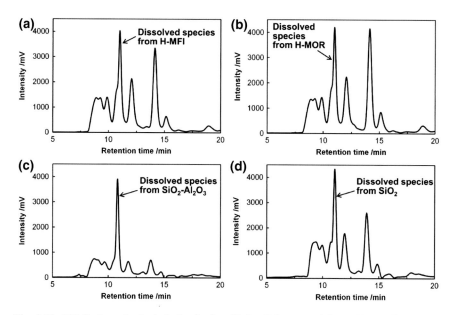

Fig. 2.10 HPLC charts for hydrolysis of mix-milled cellulose containing **a** H-MFI, **b** H-MOR, **c** SiO$_2$-Al$_2$O$_3$, and **d** SiO$_2$. Column: Rezex RPM-Monosaccharide Pb++. Detector: RI. Reaction conditions: mix-milled cellulose 374 mg (containing cellulose 324 mg and catalyst 50 mg); distilled water 40 mL; 453 K; 20 min

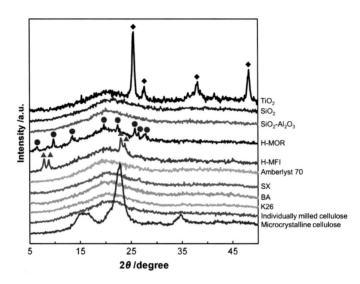

Fig. 2.11 XRD patterns of mix-milled samples containing cellulose and solid catalysts. The peaks marked with *black diamonds*, *red circles*, and *blue triangles* are from TiO_2, H-MOR, and H-MFI, respectively

(pH 2.5) was utilized as a solvent instead of water. This trace amount of HCl neither corrodes common stainless steel reactors nor has a negative economic impact because of very low concentration and low price [0.07–0.1 USD kg^{-1} as concentrated HCl in 2014 according to International Chemical Information Service (ICIS)] [34, 35]. Figure 2.12 shows the time course of the hydrolysis of mix-milled cellulose containing K26 conducted in the trace HCl aqueous solution at 453 K. In the initial stage of the reaction (≤2 min), water-soluble oligosaccharides were predominantly formed. Subsequently, the yield of glucose increased up to 88 % at 20 min and the selectivity of glucose based on cellulose conversion (98 %) was 90 % (Table 2.5, entry 44). This is the best result achieved in this work and one of the highest yields of glucose ever reported in any methods. After 20 min, the yield of glucose decreased because of the decomposition. Although an H_2SO_4 aqueous solution (0.018 wt%, pH 2.5) was advantageous over HCl due to its lower price (0.05 USD kg^{-1} in 2009 according to Nexant [36]) and less corrosive property, H_2SO_4 provided by 19 % lower yield of glucose than HCl at the same pH (entry 45). This result was possibly caused by the negative effect of SO_4^{2-} species on the hydrolysis reaction (Table 2.6); SO_4^{2-} species may prevent the activation of glycosidic bonds by forming a chelate. Therefore, the 0.012 wt% of HCl aqueous solution was employed as a reaction solvent in the following study. Note that such high-yielding production of glucose was unable to be achieved without the mix-milling pretreatment. Two types of control experiments were conducted as follows: the hydrolysis of individually ball-milled cellulose in the trace HCl (entry 46) and that of individually ball-milled cellulose by individually ball-milled K26 in

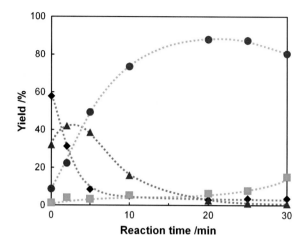

Fig. 2.12   Time course of hydrolysis of mix-milled cellulose containing K26 in 0.012 wt% HCl solution at 453 K. Conditions: mix-milled cellulose 374 mg (cellulose 324 mg and K26 50 mg); 0.012 wt% HCl solution 40 mL. Legends: *black diamonds* cellulose; *red circles* glucose; *blue triangles* oligosaccharides; and *green squares* by-products. In figure, the data sets at 0 min are for the hydrolysis under rapid heating–cooling condition (see Fig. 2.3). *Dashed lines* are simply smooth lines for connecting experimental data

Table 2.6   Inhibitory effect of salts on hydrolysis of cellulose

Entry	Salt	Conv./%	Yield based on carbon/%						
			Glucan		By-product				
			Glc[a]	Olg[b]	Frc[c]	Man[d]	Lev[e]	HMF[f]	Others[g]
1	None	28	4.6	15	0.5	0.6	0.2	1.8	4.8
49	3.2 mM NaCl	15	1.7	5.6	0.4	0.7	0.1	1.0	1.2
50	1.6 mM Na$_2$SO$_4$	6.7	0.3	2.3	0.1	0.1	0.4	0.2	3.4

Conditions individually ball-milled cellulose 324 mg (0.9 L pot); salt solution 40 mL; 503 K; rapid heating–cooling condition (Fig. 2.3)
[a]Glucose
[b]Water-soluble oligosaccharides (DP = mainly 2–6)
[c]Fructose
[d]Mannose
[e]Levoglucosan
[f]5-HMF
[g](Conversion) – (Total yield of identified products)

the HCl aqueous solution (entry 47). These control reactions yielded glucose in 20 and 30 %, which were significantly lower than the highest yield (88 %, entry 44). Similarly, the values of cellulose conversion were only 39 and 40 % in these control reactions. Neither the trace HCl nor individually ball-milled K26 accelerated the hydrolysis of cellulose in good performance due to low H$^+$ concentration and loose contact with the substrate. The mix-milling pretreatment is clearly essential for high-yielding production of glucose from cellulose.

The durability of K26 in this system was evaluated as the following procedure (see Sect. 2.2.7 as well as the schemes in Figs. 2.13 and 2.14): microcrystalline cellulose and K26 were milled together (i.e., mix-milling pretreatment); the hydrolysis of mix-milled cellulose in the 0.012 wt% HCl aqueous solution was operated at high loadings of the solid (cellulose 200 g L^{-1} and K26 31 g L^{-1}); the solid residue containing used K26 and unreacted cellulose was recovered, washed with distilled water repeatedly, and dried under vacuum; the resulting solid was ball-milled with fresh microcrystalline cellulose, the amount of which was equivalent to that of consumed cellulose in the previous run to maintain an S/C ratio; and the next run was conducted. This cycle was repeated several times. Figure 2.13 represents the results of reuse experiments at high cellulose conversion (88–91 %), and glucose was obtained in 71, 72, 70, and 67 % yields. Although these values were lower than the best glucose yield in this work (88 % at 8.1 g L^{-1}, Table 2.5, entry 44) probably due to the significantly high loadings of the solid, the high concentration of glucose clears the practical demand (10 wt%) to avoid energy-consuming condensation process of reaction product. Another type of reuse experiments at lower cellulose conversion was also performed (Fig. 2.14), in which yields of products and cellulose conversion gradually increased. This trend is reasonable since the residual solid cellulose has partially undergone hydrolysis to

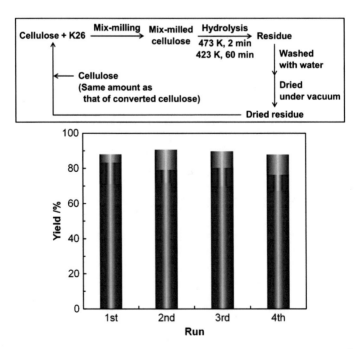

Fig. 2.13 Reuse experiments of mix-milled K26 for hydrolysis of cellulose (200 g L^{-1}) in 0.012 wt% HCl solution with high conversion. Legends: *red bar* glucose; *blue bar* oligosaccharides; and *gray bar* by-products

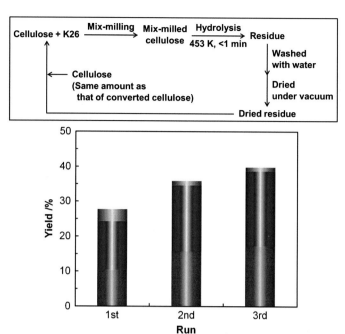

Fig. 2.14 Reuse experiments of mix-milled K26 for hydrolysis of cellulose (200 g L^{-1}) in 0.012 wt% HCl solution with low conversion. Legends: *red bar* glucose; *blue bar* oligosaccharides; and *gray bar* by-products

decrease DP in the previous run, which make this remaining part more reactive in the subsequent runs. These results in both high and low conversions show that the degradation rate of cellulose does not decrease during the repeated reactions. Hence, K26 is fairly stable during the mix-milling pretreatment as well as the hydrolysis in the presence of trace HCl and high concentrations of products.

Toward practical applications, the author demonstrated the saccharification of bagasse kraft pulp as a real biomass substrate by the inexpensive and commercially available steam-activated carbon BA50, which showed catalytic activity in cellulose conversion (Table 2.5, entry 36). The pulp contained cellulose (59 wt%), hemi-cellulose [xylan (25 wt%) and arabinan (2 wt%)], and lignin (9 wt%). At first, one-pot hydrolysis ofboth cellulose and hemicellulose in the bagasse kraft pulp was performed in the 0.012 wt% HCl solution for 2 min at various temperatures (Fig. 2.15) after mix-milling pretreatment. The yields of xylose andarabinose based on the amount of hemicellulose in the pulp were 89 and 5.2 % at 473 K, respec-tively. In this case, glucose yield based on the amount of cellulose in the pulp was only 32 %. The glucose yield increased up to 72 % with increasing the reaction temperature to 493 K, whereas the xylose yield decreased to 45 % due to the degradation. One-pot high-yielding production of both hexoses and pentoses is

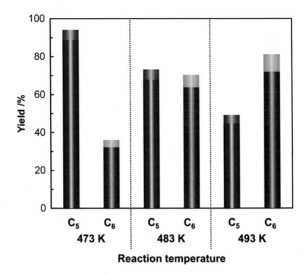

Fig. 2.15 One-pot hydrolysis of both cellulose/hemicellulose in bagasse kraft pulp at various temperatures for 2 min. Conditions: mix-milled bagasse craft pulp 374 mg (pulp 324 mg and BA50 50 mg); 0.012 wt% HCl solution 40 mL. Legends: *blue bar* xylose; *pink bar* arabinose; *red bar* glucose; and *green bar* fructose and mannose

difficult. Xylan is known as a more reactive substrate than cellulose due to the lack of an OH group at C6, which would make cellulose more chemically stable by forming hydrogen bonds (Fig. 2.16) [37]. In fact,xylobiose (the smallest model of xylan) was five times more reactive for the hydrolysis than cellobiose (that of cellulose) at 443 K for 10 min (Fig. 2.17). Taking account of the different reactivity of hemicellulose andcellulose, a cascade reaction was applied; the hydrolysis at 453 K for 1 min and subsequently at 483 K for 2 min (see the scheme in Fig. 2.18). In the first step, the hydrolysis of hemicellulose was preferred tothat of cellulose due to higher reactivity of hemicellulose, and the major products were pentoses, namely xylose (69 %) and arabinose (6.7 %). The yield of hexoses was only 1.9 % in this

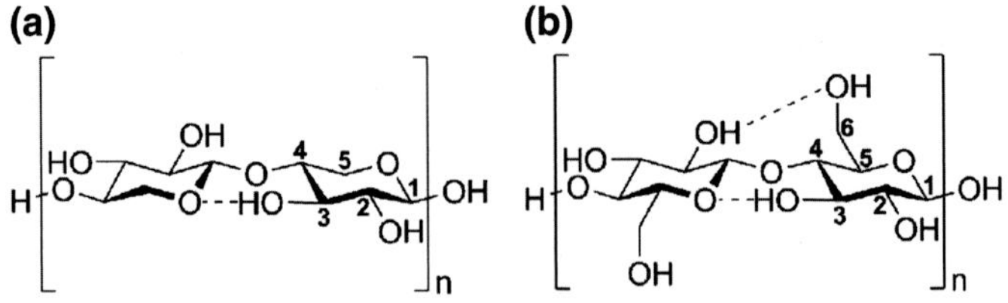

(a)   **(b)**

Fig. 2.16 Structures of **a** xylan and **b** cellulose. The hydrogen bond at OH group coordinating to C6 would make cellulose less reactive

Fig. 2.17 Hydrolysis of xylobiose (the smallest model of xylan) and cellobiose (that of cellulose) by K26 at 443 K for 10 min. Legends: *blue bar* xylose and *red bar* glucose

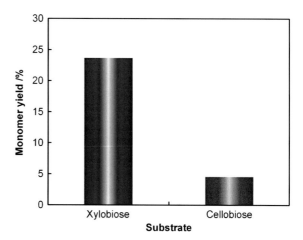

Fig. 2.18 Two-step hydrolysis of cellulose/hemicellulose in bagasse kraft pulp. Legends: *blue bar* xylose; *pink bar* arabinose; *red bar* glucose; and *green bar* fructose and mannose

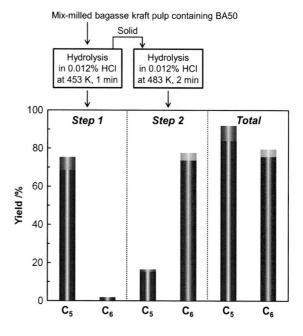

step. In contrast, glucose(74 %) and other hexoses [fructose (2.5 %) and mannose (1.5 %)] from cellulose were mainly produced in the second step. The yields of xylose and arabinose in this step were only 15 % and 1.4 %, respectively.This separate production of sugars will be advantageous for their further use. The total yields of sugars in this cascade reaction were 76 % for glucose, 4.0 % for other hexoses, 84 % for xylose, and 8.1 % for arabinose.

2.3.3 Role of Mix-Milling Pretreatment in Cellulose Hydrolysis

In order to clarify the role of the mix-milling pretreatment, individually ball-milled cellulose and mix-milled cellulose containing K26 were analyzed by microscopes (optical microscope and SEM, Fig. 2.19). The particles of individually ball-milled cellulose were clear or white and their diameter was *ca.* 10 μm (Fig. 2.19a). Individually ball-milled K26 had black small particles (≤1 μm) (Fig. 2.19b). The particle size of mix-milled cellulose containing K26 was similar to that of individually ball-milled cellulose, whereas the surface color was black (Fig. 2.19c). The particle size distributions of both individually ball-milled and mix-milled celluloses estimated by laser diffraction were also almost the same (37–40 μm, Fig. 2.20). These results suggest that small K26 particles are attached to the surface of large cellulose particles. In fact, the SEM image of the mix-milled cellulose sample (Fig. 2.19d) showed that there were small particles on a large particle. Hence, cellulose and K26 in the mix-milled sample would have a tight contact.

The author further analyzed the mix-milled cellulose containing K26 by ^{13}C CP/MAS NMR spectroscopy and viscometry to investigate chemical properties of cellulose. As well as XRD measurement (Fig. 2.11), ^{13}C CP/MAS NMR spectroscopy indicated that both individually ball-milled and mix-milled cellulose samples were in amorphous form (Fig. 2.21). The *CrI* values determined from the

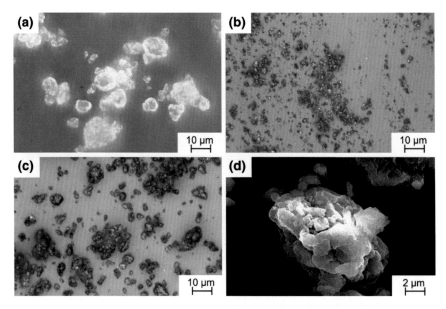

Fig. 2.19 Microscopic images of cellulose samples. Optical microscopic images of **a** individually ball-milled cellulose, **b** individually ball-milled K26, and **c** mix-milled cellulose containing K26 and **d** SEM image of the mix-milled cellulose sample

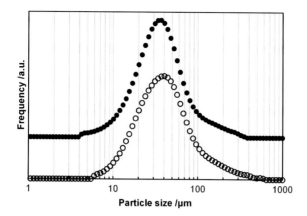

Fig. 2.20 Particle size distributions of mix-milled cellulose containing K26 (*closed circles*) and simple mixture of individually ball-milled cellulose and individually ball-milled K26 (*open circles*), which were observed by laser diffraction

Fig. 2.21 ^{13}C CP/MAS NMR spectra of mix-milled cellulose containing K26 and individually ball-milled cellulose. C4$_{cr}$ and C6$_{cr}$ show the peaks derived from crystalline cellulose and C4$_{am}$ and C6$_{am}$ represent those from amorphous one

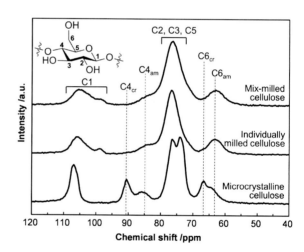

C4 signals based on Eq. 1.2 were less than 5 % for both samples. The DP of mix-milled cellulose was determined to be 690 by viscometry with a 9 wt% LiCl/DMAc solvent at 303 K and the Mark-Houwink-Sakurada equation (Eq. 2.4) [23].

$$[\eta] = KM_w^a \qquad (2.4)$$

in which $[\eta]$ (dL g^{-1}) and M_w (g mol^{-1}) are intrinsic viscosity and weight average molecular weight, respectively. The values of K and a for cellulose in the 9 wt% LiCl/DMAc solvent at 303 K have been reported as 1.278×10^{-3} dL g^{-1} and 1.19

in a reference [23]. The value of $[\eta]$ is determined from Eq. 2.5 and experimental data (Fig. 2.22).

$$[\eta] = \lim_{c \to 0} \frac{\frac{t}{t_0} - 1}{c} \qquad (2.5)$$

where c (unit: g dL^{-1}) is concentration of cellulose, t (s) is flow time of solution in a viscometer, and t_0 (s) is flow time of solvent (in this study, 9 wt% LiCl/DMAc).

The DP of mix-milled cellulose (690) was half as high as that of microcrystalline cellulose (1240), and the value was similar to that of individually milled cellulose (640). Therefore, K26 did not hydrolyze cellulose into soluble fractions during the milling process, which stood in stark contrast to H_2SO_4 [31, 33] and Amberlyst 70 (vide supra). The difference in the catalytic performance between individually ball-milled cellulose (Table 2.5, entry 26) and mix-milled cellulose (entry 35) is not ascribed to the nature of cellulose, but to the contact between the catalyst and cellulose, i.e., loose/tight contact.

To confirm the role of mix-milling pretreatment to make a tight contact between solids, the author investigated two types of model reactions: (i) cellobiose (water-soluble substrate) and K26 (insoluble catalyst) and (ii) cellulose (insoluble substrate) and benzoic acid (soluble catalyst). Even though the mix-milling pretreatment makes a tight contact between these compounds, the soluble substrate or catalyst dissolves in water to give liquid-solid reactions; in brief, the promotional effect of mix-milling should not be observed in these model reactions. Note that benzoic acid was suitable for this study employing solid-solid mixing since benzoic acid itself is solid and is a model of weakly acidic species on carbon catalysts (see Sect. 3.3.2). A typical soluble catalyst H_2SO_4 is unable to use in this model reaction, as H_2SO_4 itself is liquid and depolymerizes cellulose during the

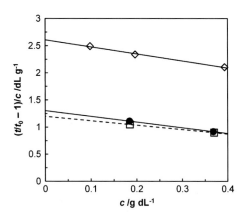

Fig. 2.22 Plots of $(t/t_{0-1})/c$ against c. In figure, the intercept is corresponding to $[\eta]$. Legends: *open diamonds* with *solid line* microcrystalline cellulose; *closed circles* with *solid line* mix-milled cellulose; and *open squares* with *dashed line* individually ball-milled cellulose

mix-milling pretreatment [31, 33]. For the combination (i), regardless of the presence/absence of mix-milling pretreatment, the hydrolysis of cellobiose by K26 proceeded at almost the same rate (Table 2.7, entries 51 and 52). Likewise, the combination (ii) showed no positive effect of mix-milling pretreatment on the hydrolysis of cellulose by benzoic acid (entries 53 and 54). In sharp contrast to these model reactions, the mix-milling pretreatment drastically accelerated the hydrolysis of solid cellulose by solid K26 and afforded *ca.* seven times higher glucan yield than individual ball-milling did (entries 35 and 55). These results suggested that the mix-milling pretreatment enhances only solid-solid reactions by forming a tight contact, but does not liquid-solid ones.

A kinetic study of the hydrolysis of cellulose was also performed to quantitatively elucidate the role of mix-milling pretreatment, in which the author chose 418 K as a reaction temperature to accurately determine kinetic parameters, as the reaction proceeds too rapidly to estimate the parameters at higher reaction temperatures such as 453 K (see Table 2.5, entry 35). Figure 2.23 shows the time course of the hydrolysis of mix-milled cellulose containing K26 in distilled water at 418 K. The amount of unreacted cellulose (black diamonds) gradually decreased with increasing reaction time. In the initial stage of the reaction, oligosaccharides (blue squares) were predominantly produced and their total yield was maximized at 6 h (44 %). After 6 h, glucose (red circles) was the major product instead of oligosaccharides. The yield of glucose increased with decrease of the yield of oligosaccharides and reached 72 % at 24 h with 97 % conversion of cellulose (Table 2.5, entry 48). This time course clearly indicates that oligosaccharides are intermediates in the reaction route from cellulose to glucose. The decomposition of

Table 2.7 Effect of solubility of substrate and catalyst on hydrolysis after mix-milling pretreatment

Entry	Pretreatment	Substrate	Catalyst	Conv./%	Yield based on carbon/%	
					Glc[a]	Olg[b]
51[c]	Milling only K26	Cellobiose	K26	12[f]	9.0	–
52[c]	Mix-milling	Cellobiose	K26	14[f]	11	–
53[d]	Milling only cellulose	Cellulose	Benzoic acid	17	3.4	9.8
54[d]	Mix-milling	Cellulose	Benzoic acid	13	2.7	8.7
55[e]	Individual milling	Cellulose	K26	18	2.9	10
35[e]	Mix-milling	Cellulose	K26	93	20	70

[a]Glucose
[b]Water-soluble oligosaccharides (DP = mainly 2–6)
[c]Conditions: cellobiose 342 mg; K26 50 mg; distilled water 40 mL; 463 K; rapid heating-cooling condition (Fig. 2.3)
[d]Conditions: cellulose 324 mg; benzoic acid 50 mg; distilled water 40 mL; 453 K; 20 min
[e]Conditions: cellulose 324 mg; K26 50 mg; distilled water 40 mL; 453 K; 20 min
[f]Conversion of cellobiose was calculated from the total amount of recovered and adsorbed cellobiose. For the estimation of adsorbed amount of cellobiose on K26, the adsorption equilibrium constant and adsorption capacity were used (see Sect. 3.3.3)

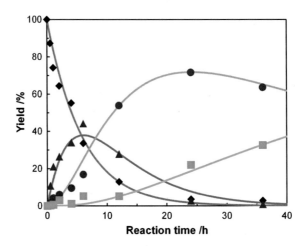

Fig. 2.23 Time course of hydrolysis of mix-milled cellulose containing K26 at 418 K. The *dots* show the experimental data and the *lines* are the results of kinetic simulations based on Eqs. 2.11–2.14. Legends: *black diamonds* cellulose; *red circles* glucose; *blue triangles* oligosaccharides; and *green squares* by-products

glucose happened to generate by-products (green triangles) after accumulation of glucose in the reaction. For these results, the hydrolysis of cellulose mainly consists of three steps as shown in Eq. 2.6 [38]: hydrolysis of cellulose to water-soluble oligosaccharides; that of oligosaccharides to glucose; and decomposition of glucose.

$$\text{Cellulose} \longrightarrow \text{Oligosaccharides} \longrightarrow \text{Glucose} \longrightarrow \text{By-products} \qquad (2.6)$$

where k_1, k_2, and k_3 (unit: h^{-1}) are pseudo first-order rate constants for hydrolysis of cellulose, that of oligosaccharides, and decomposition of glucose, respectively.

The reaction rates for respective steps are represented as Eqs. 2.7–2.10, for which first-order dependence on substrates has been postulated as reported elsewhere [39–41].

$$\frac{d[\text{Cellulose}]}{dt} = -k_1[\text{Cellulose}] \qquad (2.7)$$

$$\frac{d[\text{Oligosaccharides}]}{dt} = k_1[\text{Cellulose}] - k_2[\text{Oligosaccharides}] \qquad (2.8)$$

$$\frac{d[\text{Glucose}]}{dt} = k_2[\text{Oligosaccharides}] - k_3[\text{Glucose}] \qquad (2.9)$$

$$\frac{d[\text{By - products}]}{dt} = k_3[\text{Glucose}] \qquad (2.10)$$

where [Cellulose], [Oligosaccharides], [Glucose], and [By-products] (unit: M) are concentrations of respective compounds and t (h) is reaction time. The integration of these formulae gives Eqs. 2.11–2.14.

$$[\text{Cellulose}] = [\text{Cellulose}]_0 e^{-k_1 t} \tag{2.11}$$

$$[\text{Oligosaccharides}] = [\text{Cellulose}]_0 \frac{k_1}{k_2 - k_1} \left(e^{-k_1 t} - e^{-k_2 t} \right) \tag{2.12}$$

$$[\text{Glucose}] = [\text{Cellulose}]_0 \frac{k_1 k_2}{k_2 - k_1} \left\{ \frac{1}{k_3 - k_1} \left(e^{-k_1 t} - e^{-k_3 t} \right) + \frac{1}{k_3 - k_2} \left(e^{-k_3 t} - e^{-k_2 t} \right) \right\} \tag{2.13}$$

$$[\text{By - products}] = [\text{Cellulose}]_0 - [\text{Cellulose}] - [\text{Oligosaccharides}] - [\text{Glucose}] \tag{2.14}$$

where [Cellulose]$_0$ (unit: M) is initial concentration of cellulose and $k_1 \neq k_2 \neq k_3$.

Based on Eqs. 2.11–2.14, the author simulated the time course of the hydrolysis of mix-milled cellulose containing K26. Each simulated curve (line) in Fig. 2.23 fitted well with experimental data (dots), showing the reliability of this curve fitting. The rate constants estimated from the simulation were $k_1 = 0.17$ h^{-1}, $k_2 = 0.16$ h^{-1}, and $k_3 = 0.017$ h^{-1}. Surprisingly, the hydrolysis of cellulose to oligosaccharides proceeds as fast as that of oligosaccharides to glucose ($k_1/k_2 = 1.1$) in this reaction system, as the rate-determining step for glucose production is generally the depolymerization of cellulose to oligosaccharides due to insolubility and low reactivity of cellulose (i.e., $k_1/k_2 < 1$) [38]. The decomposition of glucose was approximately 10 times slower than the two hydrolysis reactions ($k_1/k_3 = 10$, $k_2/k_3 = 9.4$), resulting in high yields of glucose.

For quantitative comparison with the mix-milling pretreatment, the same analytic approach based on Eqs. 2.11–2.14 was performed for the hydrolysis of individually ball-milled cellulose by K26 at 418 K, in which a sampling method was employed to rigorously quantify the reaction products due to low reaction rates. Note that the reactivity of individually ball-milled cellulose itself should be similar to that of mix-milled cellulose (vide supra). In the hydrolysis of individually ball-milled cellulose, similar to that of mix-milled cellulose, the predominant formation of oligosaccharides was observed in the initial stage, followed by hydrolysis to glucose and successive degradation of glucose. The rate constants were determined to be $k_1 = 0.013$ h^{-1}, $k_2 = 0.16$ h^{-1}, and $k_3 = 0.017$ h^{-1} by the simulation. The value of k_1 was 13-fold diminished by changing the pretreatment method from mix-milling to individual ball-milling, while the values of k_2 and k_3 were the same regardless of the pretreatment methods. Accordingly, the ratio of k_1/k_2 in this case was only 0.018, exhibiting that the rate-determining step for glucose formation was the hydrolysis of solid cellulose to soluble oligosaccharides. Therefore, the author

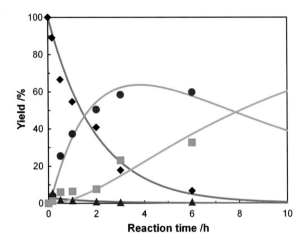

Fig. 2.24 Time course of hydrolysis of individually ball-milled cellulose by H_2SO_4 (50 mM) at 418 K. The *dots* show the experimental data and the *lines* are the results of kinetic simulations based on Eqs. 2.11–2.14. Legends: *black diamonds* cellulose; *red circles* glucose; *blue triangles* oligosaccharides; and *green squares* by-products

has quantitatively revealed that the mix-milling pretreatment selectively and drastically accelerated the solid-solid reaction.

A typical soluble acid catalyst H_2SO_4 was also tested in the hydrolysis of individually ball-milled cellulose in order to compare solid–solid and solid–liquid reactions (Fig. 2.24). In this test, 50 mM (0.49 wt%) of H_2SO_4 has been employed, as this concentration is a usual value in diluted H_2SO_4 processes [9, 10, 42]. The kinetic simulation provided the rate constants, i.e., $k_1 = 0.5$ h^{-1}, $k_2 = 17$ h^{-1}, and $k_3 = 0.12$ h^{-1}. As previously reported [38], the hydrolysis of solid cellulose to soluble oligosaccharides was significantly slower than that of oligosaccharides ($k_1/k_2 = 0.029$). Consequently, the high ratio of k_1/k_2 provided by mix-milling cellulose with K26 is specific to this new pretreatment method to form a tight solid–solid contact.

2.4   Conclusions

Simple carbon materials catalyze hydrolysis of cellulose, and their catalytic activities correlate with the amounts of weakly acidic OFGs. Such weak acid catalysts can survive even in the presence of buffer and salt potentially derived from real biomass. However, at issue is a loose contact between the solid substrate and solid catalyst, limiting the yield of glucose at most 36 % even when using the most active catalyst, the alkali-activated carbon K26. This obstacle has been overcome by the new pretreatment method, namely mix-milling cellulose and K26 to make a tight solid–solid contact. The hydrolysis of mix-milled cellulose provides 72 % yield of glucose in distilled water and 88 % yield with 90 % selectivity in the trace HCl solution, which is one of the highest glucose yields ever reported. The mix-milling pretreatment is also effective for the depolymerization of cellulose/hemicellulose in bagasse kraft pulp, resulting in 80 and 92 % yields of hexoses and pentoses, respectively. Characterization of mix-milled cellulose, a series of model reactions, and kinetic studies clearly show that the hydrolysis of water-insoluble cellulose to soluble oligosaccharides over the carbon catalyst is selectively and drastically accelerated owing to the tight contact between solid cellulose and solid K26 created by the mix-milling pretreatment.

Based on this study, Fukuoka et al. have designed an attractive recycling system for saccharification of raw biomass [43]. In their work, a carbon catalyst named *E-Carbon* has been prepared by carbonizing and oxidizing *Eucalyptus* under air and has been used for mix-milling pretreatment with *Eucalyptus*, followed by hydrolysis; that is, *Eucalyptus* is a substrate as well as a carbon source in this process. The solid residue after the reaction contain both *E-Carbon* and remaining substrate and can be utilized as a carbon catalyst again after carbonization and oxidation, making this process practicable to build up sustainable societies.

Also, it has been reported that the mix-milling pretreatment is effective for hydrolytic hydrogenation of cellulose to sorbitol by carbon-supported metal catalysts since the first step is the same hydrolysis of insoluble cellulose to soluble oligosaccharides [44, 45]. Mix-milling is a promising method to overcome a fundamental problem of solid–solid reactions, namely loose contact between solids, by making tight contact. This pretreatment method is expected to be applicable to other solid–solid reactions, e.g., oxidation of diesel soot particulates over solid catalysts [46].

References

1. Rinaldi R, Schüth F (2009) Design of solid catalysts for the conversion of biomass. Energy Environ Sci 2(6):610–626
2. Alonso DM, Bond JQ, Dumesic JA (2010) Catalytic conversion of biomass to biofuels. Green Chem 12(9):1493–1513
3. Gallezot P (2012) Conversion of biomass to selected chemical products. Chem Soc Rev 41 (4):1538–1558
4. Tuck CO, Pérez E, Horváth IT, Sheldon RA, Poliakoff M (2012) Valorization of biomass: deriving more value from waste. Science 337(6095):695–699
5. Besson M, Gallezot P, Pinel C (2013) Conversion of biomass into chemicals over metal catalysts. Chem Rev 114(3):1827–1870
6. Yabushita M, Kobayashi H, Fukuoka A (2014) Catalytic transformation of cellulose into platform chemicals. Appl Catal B Environ 145:1–9
7. Corma A, Iborra S, Velty A (2007) Chemical routes for the transformation of biomass into chemicals. Chem Rev 107(6):2411–2502
8. Serrano-Ruiz JC, West RM, Dumesic JA (2010) Catalytic conversion of renewable biomass resources to fuels and chemicals. Annu Rev Chem Biomol Eng 1:79–100
9. Kobayashi H, Fukuoka A (2013) Synthesis and utilisation of sugar compounds derived from lignocellulosic biomass. Green Chem 15(7):1740–1763

10. Rinaldi R, Schüth F (2009) Acid hydrolysis of cellulose as the entry point into biorefinery schemes. ChemSusChem 2(12):1096–1107
11. Kobayashi H, Ohta H, Fukuoka A (2012) Conversion of lignocellulose into renewable chemicals by heterogeneous catalysis. Catal Sci Technol 2(5):869–883
12. Schüth F, Rinaldi R, Meine N, Käldström M, Hilgert J, Rechulski MDK (2014) Mechanocatalytic depolymerization of cellulose and raw biomass and downstream processing of the products. Catal Today 234:24–30
13. Kobayashi H, Komanoya T, Hara K, Fukuoka A (2010) Water-tolerant Mesoporous-Carbon-Supported ruthenium catalysts for the hydrolysis of cellulose to glucose. ChemSusChem 3(4):440–443
14. Komanoya T, Kobayashi H, Hara K, Chun W-J, Fukuoka A (2011) Catalysis and characterization of carbon-supported ruthenium for cellulose hydrolysis. Appl Catal A Gen 407(1):188–194
15. Suganuma S, Nakajima K, Kitano M, Yamaguchi D, Kato H, Hayashi S, Hara M (2008) Hydrolysis of cellulose by amorphous carbon bearing SO_3H, COOH, and OH groups. J Am Chem Soc 130(38):12787–12793
16. Onda A, Ochi T, Yanagisawa K (2008) Selective hydrolysis of cellulose into glucose over solid acid catalysts. Green Chem 10(10):1033–1037
17. Mo X, López DE, Suwannakarn K, Liu Y, Lotero E, Goodwin JG Jr, Lu C (2008) Activation and deactivation characteristics of sulfonated carbon catalysts. J Catal 254(2):332–338
18. Chung P-W, Charmot A, Olatunji-Ojo OA, Durkin KA, Katz A (2013) Hydrolysis catalysis of Miscanthus xylan to xylose using weak-acid surface sites. ACS Catal 4(1):302–310
19. Charmot A, Chung P-W, Katz A (2014) Catalytic hydrolysis of cellulose to glucose using weak-acid surface sites on postsynthetically modified carbon. ACS Sustainable Chem Eng 2(12):2866–2872
20. Jun S, Joo SH, Ryoo R, Kruk M, Jaroniec M, Liu Z, Ohsuna T, Terasaki O (2000) Synthesis of new, nanoporous carbon with hexagonally ordered mesostructure. J Am Chem Soc 122(43):10712–10713
21. Sluiter A, Hames B, Ruiz R, Scarlata C, Sluiter J, Templeton D, Crocker D. Determine of structural carbohydrates and lignin in biomass: Laboratory Analytical Procedures (LAP). (Online) http://www.nrel.gov/biomass/pdfs/42618.pdf Accessed 31 Oct 2015
22. Boehm H (1994) Some aspects of the surface chemistry of carbon blacks and other carbons. Carbon 32(5):759–769
23. McCormick CL, Callais PA, Hutchinson BH Jr (1985) Solution studies of cellulose in lithium chloride and N,N-dimethylacetamide. Macromolecules 18(12):2394–2401
24. Strlič M, Kolenc J, Kolar J, Pihlar B (2002) Enthalpic interactions in size exclusion chromatography of pullulan and cellulose in LiCl-N,N-dimethylacetamide. J Chromatogr A 964(1):47–54
25. Mäurer T, Müller SP, Kraushaar-Czarnetzki B (2001) Aggregation and peptization behavior of zeolite crystals in sols and suspensions. Ind Eng Chem Res 40(12):2573–2579
26. Gopalakrishnan S, Yada S, Muench J, Selvam T, Schwieger W, Sommer M, Peukert W (2007) Wet milling of H-ZSM-5 zeolite and its effects on direct oxidation of benzene to phenol. Appl Catal A Gen 327(2):132–138
27. Mitra B, Kunzru D (2008) Washcoating of different zeolites on cordierite monoliths. J Am Ceram Soc 91(1):64–70
28. Bobleter O (1994) Hydrothermal degradation of polymers derived from plants. Prog Polym Sci 19(5):797–841
29. Strachan J (1938) Solubility of cellulose in water. Nature 141:332–333
30. Fang Z, Koziński JA (2000) Phase behavior and combustion of hydrocarbon-contaminated sludge in supercritical water at pressures up to 822 MPa and temperatures to 535 °C. Proc Combust Inst 28(2):2717–2725
31. Shrotri A, Lambert LK, Tanksale A, Beltramini J (2013) Mechanical depolymerisation of acidulated cellulose: understanding the solubility of high molecular weight oligomers. Green Chem 15(10):2761–2768

32. Hick SM, Griebel C, Restrepo DT, Truitt JH, Buker EJ, Bylda C, Blair RG (2010) Mechanocatalysis for biomass-derived chemicals and fuels. Green Chem 12(3):468–474

33. Meine N, Rinaldi R, Schüth F (2012) Solvent-free catalytic depolymerization of cellulose to water-soluble oligosaccharides. ChemSusChem 5(8):1449–1454

34. Geboers J, Van de Vyver S, Carpentier K, Jacobs P, Sels B (2011) Efficient hydrolytic hydrogenation of cellulose in the presence of Ru-loaded zeolites and trace amounts of mineral acid. Chem Commun 47(19):5590–5592

35. International Chemical Information Service (ICIS). Hydrochloric acid prices, markets & analysis. (Online) http://www.icis.com/chemicals/hydrochloric-acid/ Accessed 31 Oct 2015

36. Nexant. Sulfur/sulfuric acid market analysis. (Online) http://yosemite.epa.gov/oa/eab_web_docket.nsf/Filings%20By%20Appeal%20Number/EFAE40E5FFEFBBF785257A2A0047A C45/$File/Exhibit%2052m%20to%20Revised%20Petition%20for%20Review%20...12.52m.pdf Accessed 31 Oct 2015

37. Maréchal Y, Chanzy H (2000) The hydrogen bond network in I_β cellulose as observed by infrared spectrometry. J Mol Struct 523(1):183–196

38. Abatzoglov N, Bouchard J, Chornet E, Overend RP (1986) Dilute acid depolymerization of cellulose in aqueous phase: Experimental evidence of the significant presence of soluble oligomeric intermediates. Canad J Chem Eng 64(5):781–786

39. Saeman JF (1945) Kinetics of wood saccharification—hydrolysis of cellulose and decomposition of sugars in dilute acid at high temperature. Ind Eng Chem 37(1):43–52

40. Sasaki M, Kabyemela B, Malaluan R, Hirose S, Takeda N, Adschiri T, Arai K (1998) Cellulose hydrolysis in subcritical and supercritical water. J Supercrit Fluids 13(1):261–268

41. Kobayashi H, Ito Y, Komanoya T, Hosaka Y, Dhepe PL, Kasai K, Hara K, Fukuoka A (2011) Synthesis of sugar alcohols by hydrolytic hydrogenation of cellulose over supported metal catalysts. Green Chem 13(2):326–333

42. Faith WL (1945) Development of the Scholler process in the United States. Ind Eng Chem 37 (1):9–11

43. Kobayashi H, Kaiki H, Shrotri A, Techikawara K, Fukuoka A (2016) Hydrolysis of woody biomass by biomass-derived reusable heterogeneous catalyst. Chem Sci 7(1):692–696

44. Komanoya T, Kobayashi H, Hara K, Chun W-J, Fukuoka A (2014) Kinetic study of catalytic conversion of cellulose to sugar alcohols under low-pressure hydrogen. ChemCatChem 6(1):230–236

45. Liao Y, Liu Q, Wang T, Long J, Zhang Q, Ma L, Liu Y, Li Y (2014) Promoting hydrolytic hydrogenation of cellulose to sugar alcohols by mixed ball milling of cellulose and solid acid catalyst. Energy Fuels 28(9):5778–5784

46. Stanmore BR, Brilhac JF, Gilot P (2001) The oxidation of soot: a review of experiments, mechanisms and models. Carbon 39(15):2247–2268

Chapter 3
Mechanistic Study of Cellulose Hydrolysis by Carbon Catalysts

3.1 Introduction

In Chap. 2, the author has conducted the screening of carbon catalysts for cellulose hydrolysis and has found that catalytic activity strongly depends on their nature. The preliminary study has represented that the amount of weakly acidic sites quantified by the Boehm titration [1] positively correlates with catalytic activity for glucose production from cellulose (Fig. 2.4). Surprisingly, the alkali-activated carbon K26 similarly accelerated the reaction even after exposing acetate buffer and NaCl solution (Fig. 2.5); undoubtedly, weak acids are major active sites for the reaction. However, the detailed structure of active sites and their catalysis have still remained unclear, which should be clarified for catalytic science as well as design of highly active carbon catalysts. Hence, the first objective of this chapter is to reveal the roles of weak acids on carbon catalysts by structural characterization, model reactions, kinetics, thermodynamics, and density functional theory (DFT) calculations.

The author has so far focused on the cleavage of glycosidic bonds. Meanwhile, in heterogeneous catalysis, adsorption of substrates on catalyst surface is also essential step for starting and accelerating catalytic reactions [2–4]. In the reported works by Katz et al., cellulosic molecules chemisorbed on SiO_2 and Al_2O_3 surfaces underwent hydrolysis induced by weakly acidic OH groups ($pK_a \sim 7$) [5–7]. The hydrolysis of glycosidic bonds by weakly acidic groups is intriguing since this reaction generally requires strong acids with a pK_a value less than -3 [8]. However, these oxides were unable to hydrolyze cellulose dispersed in water. These evidences exhibit that weakly acidic groups on the solids catalyze the hydrolysis of cellulose only when the substrate molecules locate close to the active sites; likewise, the adsorption of cellulosic molecules could be involved in the highly efficient catalysis of carbon materials (Fig. 3.1). Herein, the author has investigated both thermodynamics and computational calculations to understand the adsorption process.

© Springer Science+Business Media Singapore 2016
M. Yabushita, *A Study on Catalytic Conversion of Non-Food Biomass into Chemicals*, Springer Theses, DOI 10.1007/978-981-10-0332-5_3

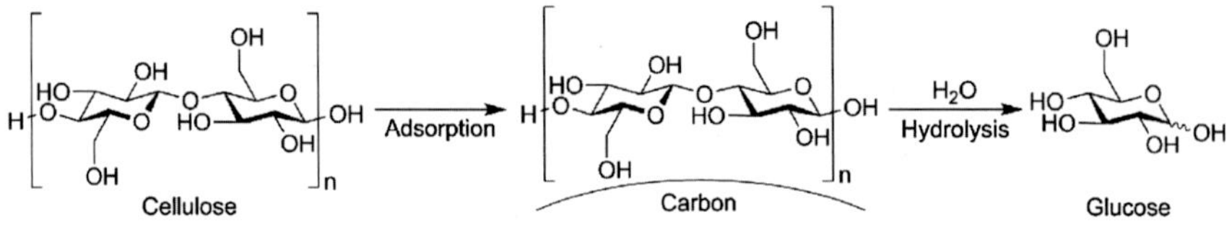

Fig. 3.1 Hydrolysis of cellulose by carbon catalysts via adsorption

3.2 Experimental

3.2.1 Reagents

Microcrystalline cellulose	Column chromatography grade, 102331, Merck
D(+)-Cellobiose	Special grade, Kanto Chemical
D(+)-Maltose	Monohydrate, special grade, Wako Pure Chemical Industries
K26	Alkali-activated carbon (not for sale), Showa Denko
ZTC	Zeolite-templated carbon [9], electrolytically oxidized, Prof. Kyotani group in Tohoku University
Phenol	Special grade, Kanto Chemical
Benzoic acid	Special grade, Wako Pure Chemical Industries
Salicylic acid	Special grade, Wako Pure Chemical Industries
m-Hydroxybenzoic acid	First grade, Wako Pure Chemical Industries
p-Hydroxybenzoic acid	First grade, Wako Pure Chemical Industries
4-Trifluoromethylsalicylic acid	Special grade, Wako Pure Chemical Industries
Phthalic acid	Special grade, Wako Pure Chemical Industries
o-Chlorobenzoic acid	Special grade, Wako Pure Chemical Industries
o-Trifluoromethylbenzoic acid	First grade, Wako Pure Chemical Industries
o-Anisic acid	First grade, Wako Pure Chemical Industries
Anthranilic acid	Special grade, Wako Pure Chemical Industries
Catechol	>99.0 %, Tokyo Chemical Industry
3,4-Dihydrocoumarin	Special grade, Wako Pure Chemical Industries
Isochroman-1-one	97 %, Wako Pure Chemical Industries
Potassium chloride	Special grade, Wako Pure Chemical Industries
Potassium bromide	Crystal block, Wako Pure Chemical Industries
Cellohexaose	>95 %, Seikagaku Biobusiness
Cellopentaose	>95 %, Seikagaku Biobusiness
Cellotetraose	>97 %, Seikagaku Biobusiness
Cellotriose	>97 %, Seikagaku Biobusiness
D(+)-Glucose	Special grade, Kanto Chemical
D(+)-Mannose	Special grade, Wako Pure Chemical Industries
D(−)-Fructose	Special grade, Kanto Chemical

1,6-Anhydro-β-D-glucopyranose	99 %, Wako Pure Chemical Industries, denoted as levoglucosan
5-Hydroxymethylfurfural	99 %, Sigma-Aldrich, denoted as 5-HMF
1-O-Methyl-α-glucose	>98.0 %, Tokyo Chemical Industry, denoted as α-MeGlc
1-O-Methyl-β-glucose	>98.0 %, hemihydrate, Tokyo Chemical Industry, denoted as β-MeGlc
Distilled water	Wako Pure Chemical Industries
Methanol	Special grade, Wako Pure Chemical Industries
Dimethyl sulfoxide	Special grade, Wako Pure Chemical Industries, denoted as DMSO
Distilled water	For HPLC, Wako Pure Chemical Industries
Deuterated dimethyl sulfoxide	For NMR, Acros Organics, denoted as DMSO-d_6
Sodium hydrogen carbonate	Special grade, Wako Pure Chemical Industries
Sodium carbonate solution	0.05 M, for volumetric analysis, Wako Pure Chemical Industries
Sodium hydroxide solution	0.05 M, for volumetric analysis, Wako Pure Chemical Industries
Hydrochloric acid solution	0.05 M, for volumetric analysis, Wako Pure Chemical Industries
Methyl orange	Special grade, Wako Pure Chemical Industries
Cyclohexane	Special grade, Wako Pure Chemical Industries
Helium gas	Alpha gas 1, Air Liquide Kogyo Gas
Helium gas	99.9999 %, Hokkaido Air Water, used for N_2 adsorption-desorption measurement
Nitrogen gas	Alpha gas 2, Air Liquide Kogyo Gas
Nitrogen gas	99.999 %, Hokkaido Air Water, used for N_2 adsorption-desorption measurement
Argon gas	Alpha gas 2, Air Liquide Kogyo Gas

3.2.2 Heat-Treatment of Alkali-Activated Carbon K26

The alkali-activated carbon K26 was heat-treated under He flow to partially remove OFGs [10] in a fixed-bed flow reactor (Fig. 3.2) at various temperatures. K26 (2.00 g) was charged into a quartz tube, and the tube was set in an electric furnace (Asahi Rika, ARF-30 K). The heater was controlled by a thyristor (Shimaden, PAC-15C, cycle calculation zero voltage switching) equipped with a program controller (Shimaden, FP93) and a thermocouple (Chino, type K, JIS class 1, ø1.6 mm) set aside of the sample. Temperature of the sample was monitored by

Fig. 3.2 Diagram of
fixed-bed flow reactor

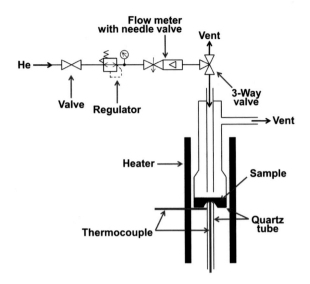

another thermocouple attached in a bottom dimple of the quartz tube. The difference in temperatures for the two thermocouples was less than 3 K in the experiments. Temperature of the heater was increased from room temperature to 393 K by 10 K min^{-1} and was maintained at 393 K for 1 h to remove physisorbed water under He flow (10 mL min^{-1}, kept by a needle bulb [Kojima Instruments (Kofloc), Variable Secondary Pressure Flow Controller Model 2203]). Then, the reactor was heated to a designated temperature (673, 873, 1073, and 1273 K) by 10 K min^{-1} under He flow (10 mL min^{-1}), and K26 was treated at the temperature for 2 h under He flow (3 mL min^{-1}). The samples treated at the temperatures were denoted as K26-673 (recovery 1.82 g), K26-873 (1.79 g), K26-1073 (1.73 g), and K26-1273 (1.72 g).

3.2.3   Characterization of Carbon Materials

Carbon materials were characterized by several physicochemical methods. The specific surface area of samples based on the Brunauer-Emmett-Teller (BET) theory was determined by N$_2$ adsorption-desorption measurement (BEL Japan, BELSORP-mini) at 77 K after drying at 393 K for 4 h under vacuum (<10 Pa) by using a pretreatment instrument (BEL Japan, BELPREP-vac II). The micropores of carbon materials were analyzed by employing another apparatus (BEL Japan, BELSORP-max, equipped with a turbo-molecular pump) at 77 K, which can detect ≥0.4 nm of micropores even using N$_2$ as an adsorbate [11]. The specific external surface area was determined by the *t*-plot method based on a standard isotherm for carbon, and pore size distribution was estimated by a nonlocal density functional theory (NLDFT) calculation (slit model of carbon black, atomic diameter of adsorbent 0.142 nm, distribution function: no assumption). The secondary particle size of carbon materials was measured in distilled water by laser diffraction (Nikkiso, Microtrac MT3300EXII).

The qualitative analysis of OFGs on carbon materials was conducted by diffuse reflectance infrared Fourier transform [DRIFT, PerkinElmer, Spectrum 100, mercury-cadmium-telluride (MCT) detector cooled down to 77 K by liquid N_2] spectroscopy. The amounts of carboxylic acids, lactones, and phenolic groups on carbons were quantified by the Boehm titration [1] with the same procedure as Sect. 2.2.3.

The framework of carbon was analyzed by solid-state ^{13}C CP/MAS-TOSS NMR (JEOL, JNM-ECX400, 100 MHz, MAS frequency 5 kHz, TOSS: total suppression of sidebands), Raman spectroscopy [Renishaw, inVia Raman microscope, Nd:YAG laser using second harmonic generation (532 nm, YAG: yttrium-aluminium-garnet)], and XRD (Rigaku, Ultima IV, Cu Kα radiation). The average size of graphene sheet in carbon was estimated using Raman spectroscopy and the Knight formula for 532 nm of laser (Eq. 3.1) [12, 13].

$$L_a = 2.4 \times 10^{-10} \lambda^4 \left(\frac{I_G}{I_D}\right)$$ (3.1)

where L_a (unit: nm) is average size of graphene sheet in carbon, λ (nm) is wavelength of laser, and I_D and I_G are intensities of D- and G-bands observed by Raman spectroscopy.

The hydrophobicity of carbon material was evaluated by cyclohexane adsorption under saturated vapor pressure at 298 K.

3.2.4 Ball-Milling Pretreatment and Hydrolysis of Cellulose

Microcrystalline cellulose (10 g) was ball-milled in the presence of ZrO_2 balls (ø1.0 cm, 1 kg) in a ceramic pot (0.9 L) at 60 rpm for 96 h, i.e., individual ball-milling. The hydrolysis of cellulose was conducted by the same procedure described in Sect. 2.2.6.

3.2.5 Hydrolysis of Cellobiose by Aromatic Compounds

Cellobiose (342 mg) and an aromatic compound aqueous solution (0.5 mM, 40 mL) were charged into the hastelloy C22 high-pressure reactor (OM Lab-Tech, MMJ-100, 100 mL, Fig. 2.2). The temperature was increased to 443 K for ca. 10 min, maintained for 10 min, and decreased to room temperature. After the reaction, products in the solution were analyzed by HPLC (see Sect. 2.2.6).

3.2.6   Analysis of Interaction Between Cellobiose and Aromatic Compounds

Both cellobiose (58 mM) and an aromatic compound (58 mM) were dissolved into DMSO-d_6 solvent, and the mixture was analyzed by ^1H NMR spectroscopy (JEOL, JNM-ECX400, 400 MHz) at 298 K.

3.2.7   Adsorption of Cellulosic Molecules

K26 (25 mg) or ball-milled cellulose (500 mg) was dispersed in an aqueous solution of cellobiose (20 mL). The suspension was stirred at 600 rpm at a designated temperature for >1 h, after which the mixture was filtered by a Mini-UniPrep [Whatman, equipped with a polyvinylidene difluoride (PVDF, 0.20 μm mesh) membrane] that was maintained at the same temperature. The filtrate was immediately analyzed by HPLC with a Rezex RPM-Monosaccharide Pb++ column (Phenomenex, ⌀7.8 × 300 mm, mobile phase: water at 0.6 mL min^{-1}, 343 K), for which an absolute calibration method was used to determine the amount of residual cellobiose in the solution. The subtraction of the amount of cellobiose detected by HPLC from that of cellobiose used in the experiments gave that of cellobiose adsorbed on solid materials. The weight of carbon used in these experiments was changed to 100 mg when studying adsorption of glucose and to 12.5 mg when using cellotriose in order to improve accuracy.

Adsorption experiment at 413 K was performed in the hastelloy-C22 high-pressure reactor into which He was introduced through a nozzle at 2 MPa (Fig. 3.3). Another nozzle was equipped with a PTFE membrane (0.10 μm mesh) on the inlet and a needle valve at the outlet. After reaching adsorption equilibrium, the solution was ejected by applying pressure of He, in which the outlet was cooled. The first 5 mL fraction was discarded and the subsequent 5 mL of solution was used for the HPLC analysis.

3.2.8   DFT Calculations

An adduct formation energy and a potential energy profile in hydrolysis were calculated using the Gaussian 09 program, which employed a cellobiose molecule, aromatic compound molecule, and water molecule. The structure of the overall system was optimized using DFT calculations at the Becke 3-parameter Lee-Yang-Parr (B3LYP) function with the 6-311G(d,p) basis set [14, 15]. The solvation effect was taken into account using the self-consistent reaction field (SCRF) method with a polarized continuum model [16] and the dielectric constant of bulk water was included.

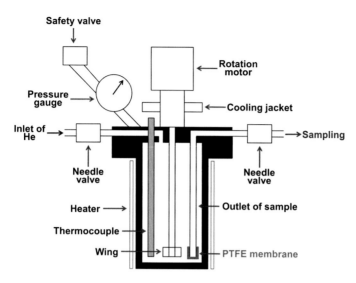

Fig. 3.3 Diagram of apparatus for adsorption experiment at 413 K. During the experiment, 2 MPa of He was charged in the apparatus

The adsorption structures of cellobiose on carbon were also calculated by the Gaussian 09 program, which employed a model carbon composed of quadruple-layered graphene sheets. The author and co-workers adopted two kinds of carbon sheets, A and B, which involved 24 and 19 six-membered rings, respectively. Sheets A and B were stacked in an alternating fashion and cellobiose was adsorbed on sheet A (see Sect. 3.3.3). The structure of the overall system was optimized by DFT calculations at the B3LYP/6-31G* level [17–19]. Dispersion interactions were also included in an empirical manner by using Grimme's D3 damping function [20]. The solvation effect was considered by the same manner as mentioned above.

3.3 Results and Discussion

3.3.1 Contributions of OFGs on Carbons for Hydrolysis of Cellulose

Hydrolysis of cellulose by various carbons indicated that the catalytic activity correlated with the amount of weakly acidic groups (Sect. 2.3.1); however, that comparison should include the influence of physical structures of respective carbons as well as that of impurities. To accurately elucidate the contributions of OFGs for the hydrolytic activity by minimizing other effects, K26 was heat-treated [10] under He flow at 673–1273 K; this treatment may provide structurally-similar carbons bearing decreased amounts of OFGs. Figure 3.4a shows the temperature course of heat-treatment for K26 against hydrolytic activity for the hydrolysis of

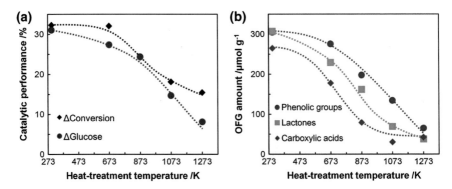

Fig. 3.4 (**a**) Hydrolysis of cellulose by heat-treated K26 catalysts and (**b**) specific amounts of OFGs. Reaction conditions: individually ball-milled cellulose (0.9 L-pot) 324 mg; catalyst 50 mg; distilled water 40 mL; 503 K; rapid heating-cooling condition (Fig. 2.3). Definitions of ΔConversion and ΔGlucose are described as Eqs. 3.2 and 3.3, respectively

individually ball-milled cellulose. In the figure, the contribution of blank reaction without catalyst (see Table 2.1) has been excluded from the values of cellulose conversion and glucose yield of catalytic reactions using heat-treated K26 samples to compare catalytic performances themselves. ΔConversion and ΔGlucose are defined as Eqs. 3.2 and 3.3.

$$
\begin{aligned}
\Delta\text{Conversion} = \ & (\text{Conversion in the catalytic reaction}) \\
& - (\text{Conversion in the blank reaction without catalyst, 28\,\%})
\end{aligned}
$$
(3.2)

$$
\begin{aligned}
\Delta\text{Glucose} = \ & (\text{Glucose yield in the catalytic reaction}) \\
& - (\text{Glucose yield in the blank reaction without catalyst, 4.6\,\%})
\end{aligned}
$$
(3.3)

Both ΔConversion and ΔGlucose gradually decreased with increasing the heat-treatment temperature, suggesting the decomposition of active sites of K26 in the heat-treatment. In fact, the drop of catalytic activity corresponded to the decrease of OFGs of heat-treated K26 samples quantified by the Boehm titration [1] (Fig. 3.4b). The amount of carboxylic acids first decreased at *ca.* 673 K, followed by the reduction of lactones. Phenolic groups were slightly more stable than lactones, which was consistent with reported results [10]. The elimination of OFGs was also observed by DRIFT spectroscopy (Fig. 3.5). In DRIFT spectra, the peak intensity at 1738 cm^{-1} derived from C=O decreased with increasing heat-treatment temperature, indicating the elimination of carboxylic acids, ketones, and lactones. This peak disappeared by heat-treatment at 1073 and 1273 K since most of C=O groups decomposed at such high temperatures. Although the intensities of the peaks due to O–H and C–O vibrations at 1241, 1403, and 3523 cm^{-1} also diminished,

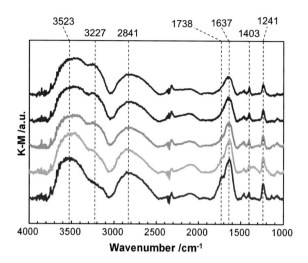

Fig. 3.5 DRIFT spectra of heat-treated K26 samples. Assignments (cm^{-1}): 3523 [ν(O–H), ν(OH–O)]; 3227 [ν(C–H, aliphatic and aromatic)]; 2841 [ν(C–H, aliphatic)]; 1738 [ν(C=O)]; 1637 [ν(C=C, aliphatic and aromatic)]; 1403 [δ(C–H, aliphatic), δ(O–H, carboxyl); γ(O–H, hydroxyl and phenol)]; 1241 [ν(C–O)]

they were still observed even after the heat-treatment at 1273 K. Hence, OFGs such as alcohols, phenols, and ethers are more thermally stable than carboxylic acids and others. These DRIFT observations are in agreement with the results of the Boehm titration. Thus, the removal of weakly acidic OFGs definitely caused the deactivation of K26 in the hydrolysis.

The textural properties of heat-treated K26 samples were also characterized by N$_2$ adsorption-desorption and laser diffraction measurements. In the N$_2$ physisorption analyses, type I isotherms were observed for all heat-treated K26 samples (Fig. 3.6), indicating their similar microporous structures. The

Fig. 3.6 N$_2$ adsorption-desorption isotherms for heat-treated K26 samples recorded at 77 K. *Solid lines* show adsorption isotherms, and *dashed lines* represent desorption ones

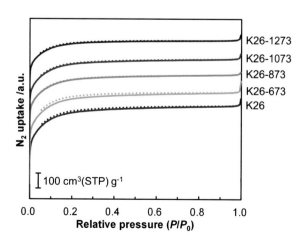

Fig. 3.7 Particle size distributions of heat-treated K26 samples in distilled water. Baselines are shifted to show respective distributions

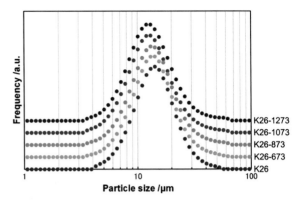

Table 3.1 Textural properties of heat-treated K26 samples

Entry	Heat-treatment temp./K	BET specific surface area/m^2 g^{-1}	Total pore volume/cm^3 g^{-1}	Median particle diameter/μm
1	None	2270	1.00	13
2	673	2240	0.99	13
3	873	1970	0.86	13
4	1073	1980	0.86	12
5	1273	1730	0.75	12

heat-treatment causes the decrease of BET specific surface area from 2270 to 1730 m^2 g^{-1} as well as total pore volume from 1.00 to 0.75 cm^3 g^{-1} (Table 3.1). The particle size distributions were not changed even after the heat-treatment (12–13 μm, Fig. 3.7 and Table 3.1). These no-large textural changes of K26 samples in the heat-treatment would not be the origin of deactivation in the hydrolysis. Instead, the OFGs should contribute to the catalytic performance of K26 samples in the hydrolysis, as carboxylic acids and phenolic groups can work as acids to activate glycosidic bonds of the substrate. Since the amounts of respective OFGs are similarly decreased in the heat-treatment, the remaining subject is to clarify the contribution of respective OFGs.

3.3.2   Clarification of Active Sites and Their Catalysis

The detailed structure of active sites and their catalysis were investigated by model reactions employing aromatic compounds bearing various OFGs as catalysts. In this study, the substrate was cellobiose, which is a minimum model of cellulose. The simplified reaction systems enabled to quantitatively evaluate the hydrolytic activities of OFGs. The results of cellobiose hydrolysis (Fig. 3.8) at 443 K are summarized in Table 3.2, in which the values of pK_a are cited from a reference [21]. TOF for glucose production was calculated from Eq. 3.4.

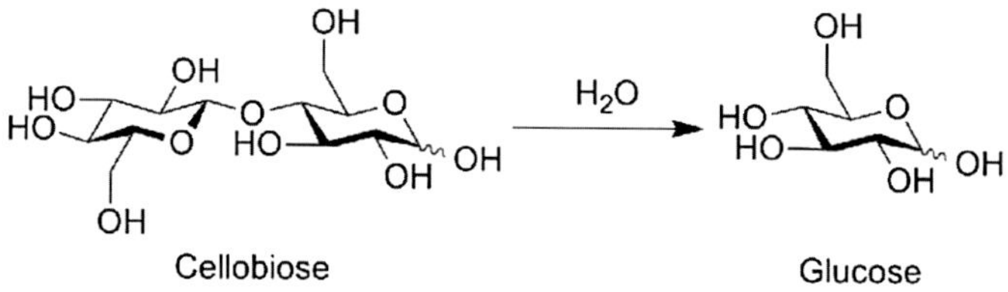

Cellobiose   Glucose

Fig. 3.8 Hydrolysis of cellobiose to glucose

Table 3.2 Hydrolysis of cellobiose by aromatic compounds

Entry	Aromatic compound	pK_a^a	TOF^b/h^{-1}	$E_a^c/kJ\ mol^{-1}$	$A^d/10^{12}\ M^{-1}\ s^{-1}$	k_H^e/s^{-1}
6	Phenol	10	0	n.d.^{f}	n.d.^{f}	43
7	Benzoic acid	4.2	4.5	111^g	1.1^g	41
8	Salicylic acid	3.0	28	118	37	190
9	m-Hydroxybenzoic acid	4.1	5.7	115^g	3.2^g	30
10	p-Hydroxybenzoic acid	4.6	2.4	115^g	2.6^g	30
11	p-Trifluoromethylsalicylic acid	2.5	46	n.d.^{f}	n.d.^{f}	$-^h$
12	Phthalic acid	3.0	32	120	73	250
13	o-Chlorobenzoic acid	2.9	17	108	2.0	23
14	o-Trifluoromethylbenzoic acid	2.6	21	n.d.^{f}	n.d.^{f}	13
15	Anisic acid	4.1	2.4	n.d.^{f}	n.d.^{f}	n.d.^{f}
16	Anthranilic acid	5.0	1.0	n.d.^{f}	n.d.^{f}	30
17	Catechol	9.5	0	n.d.^{f}	n.d.^{f}	$-^i$
18	3,4-Dihydrocoumarin	n.d.^{f}	0	n.d.^{f}	n.d.^{f}	0
19	Isochroman-1-one	n.d.^{f}	0	n.d.^{f}	n.d.^{f}	0

Conditions cellobiose 342 mg; 0.5 mM aromatic compound aqueous solution 40 mL; 443 K; 10 min
^{a}The pK_a values are cited from a reference [21]
^{b}TOF for glucose production calculated from Eq. 3.4
^{c}Activation energy
^{d}Frequency factor
^{e}Rate constant of proton-exchange determined from a peak at 6.32 ppm in ^1H NMR spectra (Fig. 3.11)
^{f}No data
^{g}1.5 mM of aromatic compound aqueous solution was used for accurate evaluation of the Arrhenius parameters
^{h}The value of k_H was not determined since the peak at 6.32 ppm was significantly broadened
^{i}The accurate evaluation of k_H was impossible due to tailing of the peak at 6.58 ppm derived from catechol

TOF

$$= \frac{(\text{Mole of glucose produced in the catalytic reaction}) - (\text{Mole of glucose produced without catalyst})}{2 \times (\text{Mole of catalyst}) \times (\text{Reaction time})}$$

$$(3.4)$$

Phenol ($pK_a = 10$) and benzoic acid ($pK_a = 4.2$) showed low catalytic activities (TOF = 0 and 4.5 h^{-1}, respectively, entries 6 and 7). o-Hydroxybenzoic acid (salicylic acid, $pK_a = 3.0$) bearing a juxtaposed carboxylic acid and phenolic group gave a high TOF of 28 h^{-1} (entry 8); in contrast, m- and p-hydroxybenzoic acids ($pK_a = 4.1$ and 4.6, respectively) were significantly less active (TOF = 5.7 and 2.4 h^{-1}, entries 9 and 10) than salicylic acid. These results imply that one or both of acidity and $ortho$ positioning of some specific groups contribute to catalytic activity for the hydrolysis. Therefore, the effect of pK_a on catalytic activity was elucidated by using 4-trifluoromethylsalicylic acids ($pK_a = 2.5$). This compound provided a very high TOF of 46 h^{-1} (entry 11), showing that a stronger acid accelerates this reaction more drastically as expected. Next, $ortho$-substituted benzoic acids were tested for evaluating the influence of vicinal functional groups. Benzene-1,2-dicarboxylic acid (phthalic acid, $pK_a = 3.0$, TOF = 32 h^{-1}, entry 12) was as effective as salicylic acid. o-Chlorobenzoic acid ($pK_a = 2.9$, TOF = 17 h^{-1}, entry 13) was less active than salicylic acid regardless of their similar acidities. o-Trifluoromethylbenzoic acid ($pK_a = 2.6$, entry 14) gave a higher TOF of 21 h^{-1} than that of o-chlorobenzoic acid due to the stronger acidity, but its catalytic performance was still less than that of salicylic acid. o-Methoxybenzoic acid (anisic acid, $pK_a = 4.1$, entry 15) and o-aminobenzoic acid (anthranilic acid, $pK_a = 5.0$, entry 16) were almost inactive. Hence, the catalytic activity also depends on the structure in addition to acidity, as salicylic acid and phthalic acid were rather more active than that expected from their pK_a values. These results suggest the synergistic effect of neighboring OH/COOH and COOH/COOH on the hydrolysis of glycosidic bonds. In a related study by Capon, the hydrolysis of a glycosidic bond is accelerated by a juxtaposed carboxylic acid 10^4 times faster than by isolated one [22]. Taking account this fact, the author has speculated that OH or COOH group interacts with cellobiose, and then another COOH activates glycosidic bonds (Fig. 3.9). Other vicinal protonic OFGs, 1,2-dihydroxybenzene (catechol, $pK_a = 9.5$), did not promote the hydrolysis at all due to its low acidity (entry 17). It is known that lactones also exist on the carbon material [10], and they potentially undergo hydrolysis to form adjacent OH/COOH during the reaction. Nevertheless, 3,4-dihydrocoumarin and isochroman-1-one were inactive for the hydrolysis of cellobiose (entries 18 and 19), showing that the effect of lactones are negligible in the hydrolysis.

In the kinetic study of this reaction, the author assumed the second-order kinetics (Eq. 3.5), and the Arrhenius equation was employed to explore the origin of high catalytic activities of salicylic acid and phthalic acid over other aromatic compounds (Fig. 3.10 and Eq. 3.6).

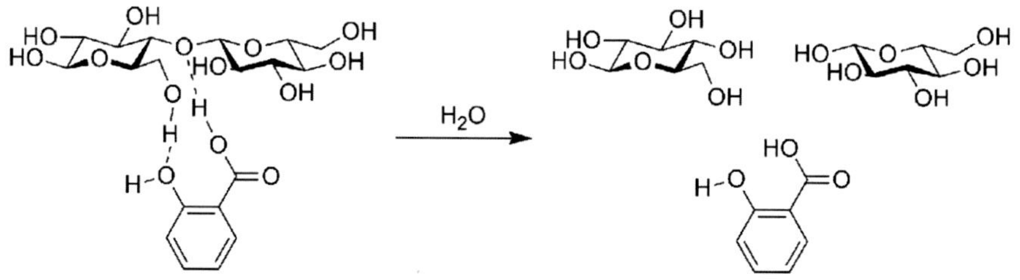

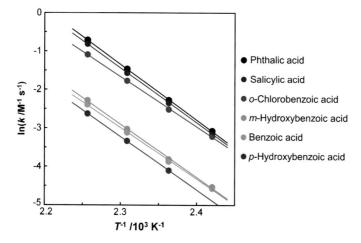

Fig. 3.9 Proposed reaction mechanism by vicinal functional groups. The detailed mechanism is depicted in Fig. 3.28

Fig. 3.10 Arrhenius plots for cellobiose hydrolysis by aromatic compounds

$$\frac{d[\text{Glucose}]}{dt} = k[\text{Cellobiose}][\text{Aromatic compound}] \qquad (3.5)$$

where [Glucose], [Cellobiose], and [Aromatic compound] (unit: M^{-1}) are concentrations of glucose, cellobiose, and aromatic compound, respectively. t (s^{-1}) is reaction time. k ($M^{-1}\ s^{-1}$) is second-order rate constant.

$$k = A\exp\left(-\frac{E_a}{RT}\right) \qquad (3.6)$$

where A (unit: $M^{-1}\ s^{-1}$) is frequency factor, E_a (J mol^{-1}) is apparent activation energy, R (J $K^{-1}\ mol^{-1}$) is the gas constant, and T (K) is reaction temperature.

The values of E_a and A for the hydrolysis of cellobiose by aromatic compounds are summarized in Table 3.2. E_a of salicylic acid and phthalic acid were 118 and 120 kJ mol^{-1}, respectively (entries 8 and 12), and other less active catalysts gave similar or rather lower E_a (108–115 kJ mol^{-1}, entries 7, 9, 10, and 13). These results are intriguing since typical catalysts accelerate reactions as a result of decline in E_a.

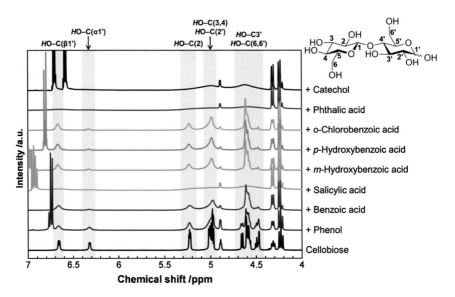

Fig. 3.11 ^{1}H NMR spectra of mixtures of cellobiose and aromatic compounds in DMSO-d$_6$. The peaks in yellow areas are derived from hydroxyl groups of cellobiose. Analytical conditions: cellobiose 58 mM; aromatic compound 58 mM; 298 K. The assignment of OH groups shown in figure is based on a reference [26]

Meanwhile, A of salicylic acid and phthalic acid (3.7×10^{13} and 7.3×10^{13} M^{-1} s^{-1}) were one-order of magnitude larger than those of other compounds (1.1×10^{12} ~ 3.2×10^{12} M^{-1} s^{-1}). Thus, the vicinal OH/COOH and COOH/COOH catalyze the hydrolysis of glycosidic bonds not by decreasing E_a, but by increasing A. It is more clearly suggested that one OH or COOH group interacts with cellobiose [23, 24] to increase the chance of glycosidic bond cleavage by a juxtaposed carboxylic acid.

To explore the interaction, ^{1}H NMR spectroscopy was conducted for cellobiose in DMSO-d$_6$ in the presence of various aromatic compounds (Fig. 3.11). Note that the signals derived from protons of hydroxyl groups can be detected in DMSO-d$_6$ because proton-exchange reaction between the solvent and solute does not happen. Both hydroxyl and CH groups of cellobiose provided sharp peaks without aromatic compound (black line). They were significantly broadened by adding salicylic acid (orange), phthalic acid (blue), and catechol (purple) due to fast proton-exchange between cellobiose and the aromatic compounds, implying the formation of hydrogen bonds [25, 26]. The rate of proton-exchange for each case was determined using a 6.32 ppm signal ascribed to a hydroxyl group coordinating to C1′ at a reducing end of cellobiose (depicted in Fig. 3.11) and Eq. 3.7 [27, 28], as this peak was isolated from other peaks of cellobiose. The proton-exchange rate constants (k_H) for salicylic acid and phthalic acid were 180 and 240 s^{-1} as represented in Table 3.2 (entries 8 and 12). Unfortunately, k_H for catechol was not accurately determined since the signal at 6.32 ppm partially overlapped with another signal at 6.58 ppm derived from catechol.

$$k_{\mathrm{H}} = \pi\left(\nu_{\mathrm{aromatic}} - \nu_{\mathrm{cellobiose}}\right) \qquad (3.7)$$

where k_{H} (unit: s^{-1}) is rate constant of proton-exchange, ν_{aromatic} (s^{-1}) is half-value width of a peak for a mixture of cellobiose and aromatic compound, and $\nu_{\mathrm{cellobiose}}$ (4.20 s^{-1}) is half-value width of a peak for only cellobiose.

Substituted benzenes bearing no vicinal OH/COOH or COOH/COOH groups such as o-chlorobenzoic acid slightly broadened the OH peaks of cellobiose. The rates of proton-exchange in these cases were one-order of magnitude lower than those of salicylic acid and phthalic acid (Table 3.2). Besides, a mixture of benzoic acid and phenol did not cause significant broadening (not shown). The fast proton exchange is not due to high acidity, but specific to vicinal OH/COOH and COOH/COOH. This result also indicates that covalently-bonded Cl species does not interact with hydroxyl groups of cellobiose in contrast to Cl$^-$ species [29, 30]. It is possible that the contamination of water in DMSO-d_6 influences on the signal broadening, but this effect is negligible since the addition of water in the solvent has not caused the broadening but done a peak shift (Fig. 3.12). Hence, vicinal OFGs uniquely interact with cellobiose by forming hydrogen bonds and subsequently exchange protons with the substrate [25, 26]. Owing to the better interaction with salicylic acid and phthalic acid, the substrate gains a larger chance to undergo activation and hydrolysis.

The interaction between cellobiose and aromatic compounds was scrutinized by DFT calculations. For the simulations, the author and co-workers assumed the association structure of cellobiose and $ortho$-substituted benzoic acid as follows: (i) a carboxylic acid makes a hydrogen bond with the O atom of glycosidic bond of cellobiose and (ii) another adjacent group does with a hydroxyl group of cellobiose. This hypothesis is based on the experimental results as shown above; that is, salicylic acid and phthalic acid both promote the hydrolysis of cellobiose by

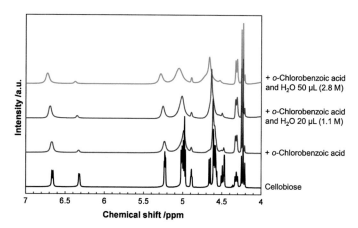

Fig. 3.12 ^{1}H NMR spectra of mixtures of cellobiose and o-chlorobenzoic acid in DMSO-d_6 in the presence of water

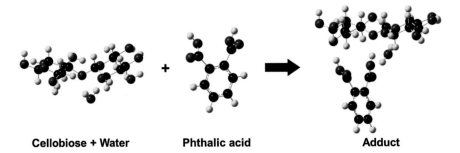

Cellobiose + Water   Phthalic acid   Adduct

Fig. 3.13 Example of association model employed for DFT calculations, in which the overall system was optimized at the B3LYP/6-311G(*d,p*) level. Legends: *black* carbon atoms; *red* oxygen atoms; and *yellow* hydrogen atoms

enhancing frequency factor. In addition, one water molecule located near cellobiose to form hydrogen bonds with cellobiose and aromatic compounds was included in the system to more accurately reproduce the configuration under the real experimental conditions. The association model for the DFT calculations is shown as Fig. 3.13. The energy of association ($E_{\text{association}}$) is estimated from Eq. 3.8.

$$E_{\text{association}} = E_{\text{adduct}} - \left(E_{\text{cellobiose + water}} + E_{\text{aromatic}}\right) \qquad (3.8)$$

where E_{adduct}, $E_{\text{cellobiose}}$, and $E_{\text{cellobiose}}$ (unit: kJ mol^{-1}) are energy of adduct, that of cellobiose and water, and that of aromatic compound, respectively.

The computation results are summarized in Table 3.3. $E_{\text{association}}$ for benzoic acid was -35 kJ mol^{-1} (entry 20), showing that the association between cellobiose and benzoic acid is energetically possible. The formation of adduct with salicylic acid (-46 kJ mol^{-1}, entry 21) and phthalic acid (-64 kJ mol^{-1}, entry 22) was more energetically favorable than that with benzoic acid. Phthalic acid provides a mono-negatively charged carboxylate species (i.e., phthalate) in water media because of its pK_a value (3.0), and this species have formed a further stable adduct in the simulation (-78 kJ mol^{-1}, entry 23). Meanwhile, $E_{\text{association}}$ for *o*-chlorobenzoic acid (-37 kJ mol^{-1}, entry 24) was almost the same as that for benzoic acid, indicating that Cl group did not play important roles in association with cellobiose. The order of stability of adduct (phthalic acid > salicylic acid > *o*-chlorobenzoic acid > benzoic acid) is the same as that of apparent frequency factor

Table 3.3 Energy of association between cellobiose and aromatic compounds, estimated by DFT calculations at the B3LYP/6-311G(*d,p*) level

Entry	Aromatic compound	$E_{\text{association}}$[a]/kJ mol^{-1}
20	Benzoic acid	-35
21	Salicylic acid	-46
22	Phthalic acid	-64
23	Phthalate[b]	-78
24	*o*-Chlorobenzoic acid	-37

[a]Energy of association, calculated from Eq. 3.8
[b]Mono-negatively charged carboxylate

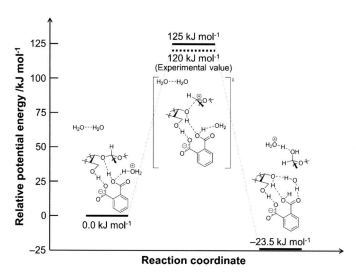

Fig. 3.14 Potential energy diagram for cellobiose hydrolysis catalyzed by mono-negatively charged phthalate species, elucidated by DFT calculations at the B3LYP/6-311G (*d,p*) level with SCRF

determined from the Arrhenius plot (vide supra); that is, these simulation results agree well with the experimental results. The author concludes that salicylic acid and phthalic acid favorably interact with cellulosic molecules by forming hydrogen bonds to enhance the activation and hydrolysis of glycosidic bonds by carboxylic acid.

The author and co-workers further pursued the reaction mechanism of glycosidic bond cleavage by the DFT calculations at the B3LYP/6-311G(*d,p*) level with SCRF (Fig. 3.14), in which a phthalate anion was employed as a catalyst since this species provided the most stable adduct in the previous estimations (vide supra). The calculation model included three water molecules; two of them stabilized the adduct and the other was a reactant in the hydrolysis. In the initial stage of the hydrolysis, a carboxylic acid of phthalate activated a glycosidic bond of cellobiose, and oxonium ion was beside phthalate to stabilize the negatively charged species. A proton of phthalate was transferred to the O atom of the glycosidic bond, resulting in the cleavage of glycosidic bond to form an oxocarbenium-like species [31] in the transition state. Similar intermediate has also been proposed in the previous reports [32, 33]. The energy of the transition state has been calculated to be 125 kJ mol^{-1} with imaginary frequency 151i cm^{-1} and is consistent with E_a estimated from the experiment (120 kJ mol^{-1}, Table 3.2, entry 12). Then, a water molecule approached the oxocarbenium species from the upper side in the figure due to the steric hindrance by a counterpart of the oxocarbenium ion and reacted to form a hydroxyl group (-23.5 kJ mol^{-1}). Accordingly, the stereochemistry of the formed hydroxyl group (i.e., α-type configuration) is opposite to the original glycosidic bond (β-1,4-glycosidic bond). This calculation result predicts that the hydrolysis of glycosidic bond proceeds as S_N1 mechanism with inversion of stereochemistry. In control experiments adding nucleophiles such as Br^- and Cl^- in cellobiose hydrolysis (Table 3.4), the reactions proceeded at the similar rates regardless of significant difference of nucleophilicity (i.e., Br^- : Cl^- : H_2O = 10000 : 1000 : 1). These results deny the possibility of S_N2 mechanism and support the predicted S_N1 mechanism by DFT calculations. Additionally, another type of control reaction, methanolysis of sugar molecules, was tested to confirm the stereochemistry of products by scission of glycosidic bonds. It is noteworthy that hydrolysis is not suitable for this clarification since reaction products, α- and β-glucose, quickly epimerize into each other in water due to the presence of hemiacetal group. In contrast, methanolysis provides 1-*O*-methyl-α-glucose and 1-*O*-methyl-β-glucose (denoted as α-MeGlc and β-MeGlc, respectively), which should suppress the epimerization owing to the absence of hemiacetal group. Table 3.5 shows the results of methanolysis of cellobiose (β-1,4-glycosidic bond) and maltose (α-1,4-glycosidic bond). The major product from cellobiose was α-MeGlc, and β-MeGlc was predominantly formed from maltose; the stereochemistry was inverted possibly due to the steric hindrance as mentioned above. These data of control experiments agree well with the computation ones, showing good reliability of the calculations.

At the end of this section, catalytic activities of salicylic acid and phthalic acid were clarified in the hydrolysis of cellulose in order to compare with that of carbon

Table 3.4 Hydrolysis of cellobiose by phthalic acid in the presence of nucleophiles

Entry	Additive	Conv./%	Yield based on carbon/%	
			Glc[a]	Others[b]
25	None	22	16	6
26	KCl[c]	26	17	9
27	KBr[c]	26	18	8
28	KBr[d]	28	19	9

Conditions cellobiose 342 mg; 0.5 mM phthalic acid aqueous solution 40 mL; 443 K; 10 min
[a]Glucose
[b](Conversion) − (Yield of glucose)
[c]25 mM
[d]100 mM

Table 3.5 Methanolysis of disaccharides

Entry	Substrate	Catalyst	Temp./K	Yield based on carbon/%	
				α-MeGlc	β-MeGlc
29	Cellobiose	Phthalic acid	473	5.5	2.5
30	Maltose	Phthalic acid	473	3.5	5.9
31	Cellobiose	K26	483	26	16

Conditions substrate 342 mg; methanol 40 mL; rapid heating-cooling condition (Fig. 2.3)

Table 3.6 Hydrolysis of cellulose by salicylic acid and phthalic acid

Entry	Catalyst	Conv./%	Yield based on carbon/%						
			Glucan		By-product				
			Glc[a]	Olg[b]	Frc[c]	Man[d]	Lev[e]	HMF[f]	Others[g]
32	None	28	4.6	15	0.5	0.6	0.2	1.8	5.3
33	Salicylic acid	37	11	19	1.1	1.0	0.6	3.2	0.9
34	Phthalic acid	39	13	20	0.9	0.8	0.7	3.1	0.9
35	K26	60	36	2.5	2.7	2.6	2.1	3.4	11

Conditions individually ball-milled cellulose (0.9 L-pot) 324 mg; catalyst 14 μmol based on mole of carboxylic acids; distilled water 40 mL; 503 K; rapid heating-cooling condition (Fig. 2.3)
[a]Glucose
[b]Water-soluble oligosaccharides (DP = mainly 2–6)
[c]Fructose
[d]Mannose
[e]Levoglucosan
[f]5-HMF
[g](Conversion) − (Total yield of the identified products)

catalyst K26 (Table 3.6). The mole of carboxylic acids was the same in the hydrolysis to fairly compare catalytic activities. Both salicylic acid and phthalic acid promoted the hydrolysis and afforded 11 and 13 % yields of glucose, respectively (entries 33 and 34). However, their catalytic activities were obviously lower than that of K26 (36 % yield, entry 35). These results suggest that K26 has other functions for cellulose hydrolysis in addition to OFGs.

3.3.3 Adsorption of Cellulosic Molecules onto K26

Considering the higher catalytic activity of alkali-activated carbon K26 than model aromatic compounds, K26 may have functions for the adsorption process in addition to those for the hydrolytic step. In Sect. 3.1, the author has already mentioned that the adsorption process of substrates on catalyst surface is crucial for heterogeneous catalysis to draw the potential [2–4]. In fact, Katz et al. has revealed that weakly acidic OH groups on the surface of SiO_2 and Al_2O_3 ($pK_a \sim 7$) can hydrolyze cellulosic molecules only when the substrate strongly interacts with the surface [5–7]. Consequently, it is necessary to study the adsorption of cellulosic molecules on K26 to understand the origin of high catalytic activities, for which glucose and water-soluble β-1,4-glucans have been used as models of cellulosic molecules in this work.

Insight into structure of carbon surface is essential to evaluate the adsorption manner, and the author characterized K26 by solid-state ^{13}C CP/MAS-TOSS NMR and Raman spectroscopy. A broad peak at 131 ppm in the ^{13}C NMR spectrum (Fig. 3.15) is derived from aromatic carbon atoms [34], indicating that the framework of K26 is mainly composed of aromatic rings. The Raman spectrum of K26

Fig. 3.15 ^{13}C
CP/MAS-TOSS NMR
spectrum of K26. The peak at
131 ppm is ascribed to
aromatic carbon atoms [34]

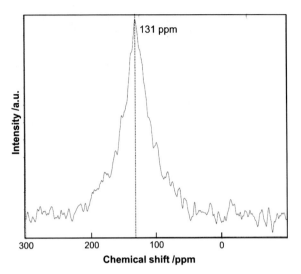

Fig. 3.16 Raman spectrum of
K26, recorded by using
532 nm of laser

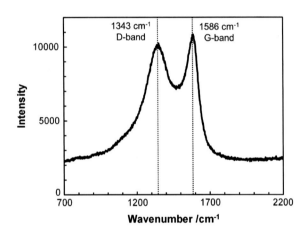

(Fig. 3.16) represents both G-band (1586 cm^{-1}) and D-band (1343 cm^{-1}) in an intensity ratio (I_G/I_D) of 1.1. An average size of graphene sheet of K26 is calculated to be 21 nm using the ratio and the Knight formula (Eq. 3.1) [12, 13].

The time course of cellobiose adsorption on K26 was measured at 296 K [Fig. 3.17, W (unit: g g$^{-1}_{adsorbent}$) in the figure represents specific adsorption amount]. 235 mg g$^{-1}_{adsorbent}$ of cellobiose adsorbed onto K26 in 1 min (the shortest time for reliable experiments), thus showing that cellulosic molecules indeed interacted with carbon surface. The adsorption amounts of cellobiose were constant regardless of the increase of contact time to 30 min, indicating that the adsorption rapidly reaches equilibrium within 1 min.

For the determination of adsorption parameters, adsorption amount of cellobiose on K26 was measured in water at 296 K by varying initial concentration of

Fig. 3.17 Time course of cellobiose adsorption on K26. Conditions: K26 25 mg; initial concentration of cellobiose 1 mM; distilled water 20 mL

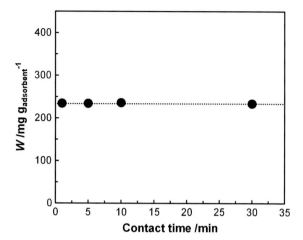

cellobiose (Fig. 3.18a, red line). It is notable that C (unit: M) in the figure does not represent the initial concentration of cellobiose but shows the concentration after reaching adsorption equilibrium. The specific adsorption amount (W) drastically increased in the range of low C of 0–4 mM; then, the isotherm became plateau above 4 mM, showing saturated adsorption. This trend is specific to a type I isotherm, and thus Langmuir-type adsorption would occur in this system. The Langmuir plot, a plot of CW^{-1} as a function of C, provided a linear line with an excellent determination coefficient ($R^2 = 0.999$) (Fig. 3.18b, red line). Monolayer adsorption of cellobiose on K26 should be predominant over multilayer one during this adsorption process since the interactions between cellulosic molecules in an aqueous solution are negligible (vide infra). Additionally, based on the theory of adsorption, monolayer adsorption is favorable in the regions of isotherms associated with a drastic increase in adsorption amount even when multilayer adsorption occurs [35].

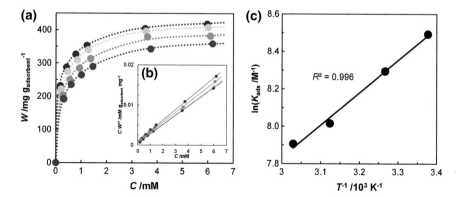

Fig. 3.18 Adsorption of cellobiose on K26: **a** isotherms, **b** Langmuir plots, and **c** a van't Hoff plot. Legends: *red* 296 K; *orange* 306 K; *green* 320 K; and *blue* 330 K

Table 3.7 Langmuir parameters for adsorption of glucose and cello-oligosaccharides on K26 at various temperatures

Entry	Adsorbate	T/K	W_{max}^a/mg $g_{adsorbent}^{-1}$	K_{ads}^b/M^{-1}
36	Glucose	296	95.2	220
37	Glucose	306	85.5	208
38	Glucose	320	76.3	172
39	Glucose	330	69.0	157
40	Cellobiose	296	412	5660
41	Cellobiose	306	404	4660
42	Cellobiose	320	387	3410
43	Cellobiose	330	360	3130
44	Cellotriose	296	527	181000
45	Cellotriose	306	517	135000
46	Cellotriose	320	501	95800
47	Cellotriose	330	485	77500

[a]Adsorption capacity
[b]Adsorption equilibrium constant

Therefore, the author remarked that cellobiose adsorbed on K26 as a Langmuir-type fashion. Similar to this conclusion, Langmuir-type adsorption isotherms were also observed in the reported works [36, 37]. Accordingly, an adsorption equilibrium constant (K_{ads}, unit: M^{-1}) and an adsorption capacity (W_{max}, unit: g $g_{adsorbent}^{-1}$) at 296 K were determined to be 5660 M^{-1} and 412 mg $g_{adsorbent}^{-1}$ (Table 3.7, entry 40) by using the Langmuir plot and the Langmuir formula (Eq. 3.9). The W_{max} means that cellobiose having a cross-sectional area of 0.8 nm^2 per one molecule occupies a specific surface area of 580 m^2 g^{-1} on K26 in the monolayer adsorption. This calculation suggests that cellobiose adsorbs in the micropores of K26 because the occupied surface area is 7.4 times greater than the specific external surface area of K26 (88 m^2 g^{-1}). The molecular size of cellobiose is 1.1 × 0.7 × 0.4 nm (Fig. 3.19). The non-uniform pores of K26 have 0.7 nm of average diameter, estimated by the t-plot method, which is similar to the minimal cross-sectional area that a cellobiose molecule requires for uptake (0.7 × 0.4 nm). Cellobiose molecules in water are solvated and the hydrated size is larger than the molecular size itself; however, the hydrated water molecules are partially removed during the adsorption process (i.e., hydrophobic interaction, vide infra). Hence, the author argues that the adsorption of cellobiose molecules occurs within micropores of K26. The further adsorption study employing a microporous carbon material has revealed that adsorbed cellulosic molecules locate in the micropores of carbon [38].

$$\frac{C}{W} = \frac{C}{W_{max}} + \frac{1}{K_{ads} W_{max}} \tag{3.9}$$

in which K_{ads} is defined by Eq. 3.10.

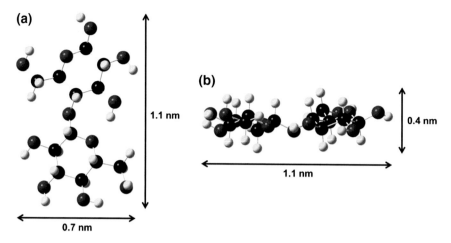

Fig. 3.19 Molecular size of cellobiose: **a** top view and **b** side view. Legends: *black* carbon atoms; *red* oxygen atoms; and *yellow* hydrogen atoms. The structure of cellobiose molecule was optimized by the DFT calculations at the B3LYP/6-31G* level

$$K_{ads} = \frac{\theta}{[Adsorbate](1 - \theta)} \tag{3.10}$$

where θ is coverage.

Regarding the adsorption of cellulosic molecules on carbon surface, two different driving forces have been postulated so far: (i) hydrogen bonds between OH groups of cellulosic molecules and OFGs of carbon [23, 24, 36] and (ii) van der Waals forces with CH–π hydrogen bonds between CH groups of axial plane of glucose units and aromatic rings of carbon material [37, 39]. To examine the former hypothesis, the author tested heat-treated K26 samples, K26-873 and K26-1273, in the control adsorption experiments at 296 K (Fig. 3.20), as these materials contain different amount of OFGs (Fig. 3.4b and Table 3.8). Although the OFG amounts of K26-873 (440 µmol g^{-1}, entry 49) and K26-1273 (150 µmol g^{-1}, entry 50) were significantly less than those of K26 (880 µmol g^{-1}, entry 48), these carbon materials had similar W_{max} values (412, 432, and 342 mg g$^{-1}_{adsorbent}$ for K26, K26-873, and K26-1273, respectively). In particular, the normalization of W_{max} by BET specific surface area provided almost the same values (0.181, 0.219, and 0.198 mg m$^{-2}_{adsorbent}$ for K26, K26-873, and K26-1273). Specific surface area roughly reflects the number of aromatic hydrophobic sites since the framework of K26 is composed of aromatic carbon atoms (Fig. 3.15) and the surface area occupied by OFGs is at most 3 %. The same trend was also observed in hydrophobicity evaluation for the K26 samples using cyclohexane as a hydrophobic probe; the adsorption amounts of cyclohexane normalized by specific surface were 0.301, 0.314, and 0.273 mg m$^{-2}_{adsorbent}$ for K26, K26-873, and K26-1273, respectively, under saturated vapor pressure. As well as W_{max}, K_{ads} values for the K26 samples were comparable to each other

Fig. 3.20 Adsorption of cellobiose on heat-treated K26 samples: **a** isotherms and **b** Langmuir plots. Conditions: cellobiose aqueous solution 20 mL; K26 25 mg. Legends: *red* K26; *green* K26-873; *blue* K26-1273

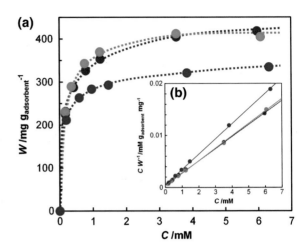

Table 3.8 Effect of OFGs amounts on Langmuir parameters for cellobiose adsorption on heat-treated K26 samples

Entry	Adsorbent	OFGs[a]/ $\mu mol \ g^{-1}$	BET[b]/ $m^2_{adsorbent} \ g^{-1}$	W^c_{max}/mg $g^{-1}_{adsorbent}$	W_{max}/BET[d]/mg $m^{-2}_{adsorbent}$	K^e_{ads}/ M^{-1}
48	K26	880	2270	412	0.181	5660
49	K26-873	440	1970	432	0.219	5690
50	K26-1273	150	1730	342	0.198	5290

[a]Total specific amount of OFGs determined by the Boehm titration [1]
[b]BET specific surface area
[c]Adsorption capacity
[d]Adsorption capacity divided by BET specific surface area
[e]Adsorption equilibrium constant

(5290–5690 M^{-1}) regardless of large difference of OFGs quantities. These results exhibit that cellobiose molecules have adsorbed on hydrophobic surface of carbon material, not on hydrophilic OFGs. This hydrophobic adsorption on carbons resemble enzymatic system; cellulases adsorb on cellulose through carbohydrate binding sites that consist of hydrophobic aromatic amino acid residues [40, 41].

To clarify the roles of hydrophobic groups, adsorption thermodynamics was studied on basic chemistry. Cellobiose was adsorbed on K26 at various temperatures, and all experiments at the temperatures provided the Langmuir-type adsorption isotherms (Fig. 3.18a, b). The adsorption parameters calculated from Eq. 3.9 were summarized in Table 3.7. K_{ads} was reduced from 5660 to 3130 M^{-1} upon increasing the adsorption temperature from 296 to 330 K (entries 40–43), exhibiting exothermic adsorption. The change of Gibbs energy for adsorption process (ΔG°_{ads}) is always calculated to be negative; for example, −21.3 kJ mol^{-1} at 296 K and −22.1 kJ mol^{-1} at 330 K, based on Eq. 3.11 and K_{ads} in Table 3.7.

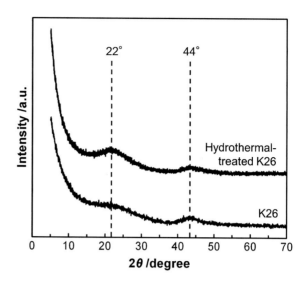

Fig. 3.21 XRD patterns of pristine K26 and hydrothermal-treated K26. The hydrothermal treatment was conducted at 453 K. The broad peaks at 22° and 44° are typically observed for activated carbons [44]

The negative $\Delta G°_{ads}$ means that the adsorption is exergonic process and thus spontaneously happens.

$$\Delta G°_{ads} = -RT\ln K_{ads} \qquad (3.11)$$

W_{max} also decreased from 412 to 360 mg $g^{-1}_{adsorbent}$ with elevating temperature. A decline in adsorption capacity at higher temperature is commonly observed in adsorption process since the adsorption cross-sectional area of adsorbate enlarges due to thermal molecular motion [42, 43]. This drop in W_{max} is not caused by decomposition of K26 in hot water, as W_{max} changes reversibly by varying temperature and the original properties of K26, the specific surface area (2270 m^2 g^{-1}) and XRD pattern (broad peaks at 22° and 44° that are typically observed for activated carbons [44], Fig. 3.21), are not changed after hydrothermal treatment even at 453 K.

For the Langmuir parameters at different temperatures, the van't Hoff equation (Eqs. 3.12 and 3.13) was employed to estimate both changes of enthalpy ($\Delta H°_{ads}$, unit: J mol^{-1}) and entropy ($\Delta S°_{ads}$, unit: J K^{-1} mol^{-1}) in adsorption process. The van't Hoff plot, a plot of $\ln K_{ads}$ against $1/T$, showed good linearity ($R^2 = 0.996$, Fig. 3.18c) due to negligible heat capacity differences. From the plot and Eq. 3.13, $\Delta H°_{ads}$ and $\Delta S°_{ads}$ for cellobiose adsorption on K26 were determined to be − 14.1 ± 1.9 kJ mol^{-1} and +23.5 ± 3.4 J K^{-1} mol^{-1}, respectively (Table 3.9, entry 52). The negative $\Delta H°_{ads}$ and positive $\Delta S°_{ads}$ values mean both enthalpically and entropically favored adsorption process.

$$\frac{d\ln K_{ads}}{d\left(\frac{1}{T}\right)} = -\frac{\Delta H°_{ads}}{R} \qquad (3.12)$$

Table 3.9 Adsorption enthalpy and entropy changes for glucose and cello-oligosaccharides on K26

Entry	Adsorbate	$\Delta H^{\circ}_{ads}/$ kJ mol^{-1}	$\Delta S^{\circ}_{ads}/$J K^{-1} mol^{-1}
51	Glucose	-8.4 ± 1.2	$+16.5 \pm 3.8$
52	Cellobiose	-14.1 ± 1.9	$+23.5 \pm 3.4$
53	Cellotriose	-20.2 ± 0.8	$+32.2 \pm 3.3$

$$\ln K_{ads} = -\frac{\Delta H^{\circ}_{ads}}{R}\frac{1}{T} + \frac{\Delta S^{\circ}_{ads}}{R} \qquad (3.13)$$

For the hydrophobic adsorption on aromatic rings (vide supra), possible interaction is CH–π hydrogen bonding between CH groups of cellobiose and π electrons of graphene sheets of K26, in addition to van der Waals forces. The ΔH°_{ads} value per mole of CH groups is -2.8 kJ mol^{-1}, which is consistent with the reported enthalpy change for the formation of CH–π hydrogen bond (-2.8 kJ mol^{-1}) [45]. The author and co-workers investigated the adsorption fashion of cellobiose on graphene sheets by the DFT calculations, in which a cellobiose molecule and quadruple-layered graphene sheets were employed. In the initial state, the cellobiose molecule was at a distance from the graphene sheets and its axial plane faced the carbon. The optimization of the whole system resulted in the adsorption of cellobiose onto the top of the graphene sheets via the axial plane (Fig. 3.22). Even when the equatorial plane of cellobiose faced the carbon in the initial placement, the cellobiose molecule spontaneously rotated and the axial planes bound with the carbon surface. As a result, the axial CH groups of cellobiose were directed toward the centers of aromatic rings of carbon, showing the formation of CH–π hydrogen bonds. For the accurate estimation of adsorption energy by DFT calculations, it is necessary to consider van der Waals forces, solvent water molecules, and upper

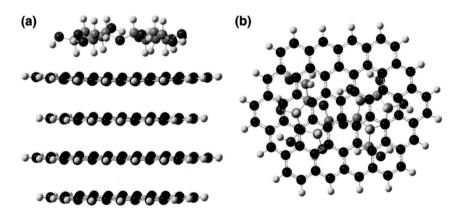

(a)   **(b)**

Fig. 3.22 Optimized adsorption orientation of cellobiose onto graphene sheets by DFT calculations: **a** side view and **b** top view. In the top view, three lower graphene sheets are omitted to assist in viewing of the structure. Legends: *black* carbon atoms of graphene sheets; *blue* carbon atoms of cellobiose; *red* oxygen atoms; and *yellow* hydrogen atoms

orbitals of hydrogen atoms (6-31G* basis set employed in this study takes account of only their s-orbitals); however, this calculation requires unfeasible computational costs. The conclusion led from the current results is that the computation qualitatively supports the formation of CH–π hydrogen bonds.

The positive ΔS°_{ads} (+23.5 J K^{-1} mol^{-1}) observed in this adsorption study is surprising since general adsorption process provides negative ΔS°_{ads} values; the degree of freedom (DF) of molecules in a 3D space is diminished by adsorption onto a 2D surface. For example, ΔS°_{ads} is −3.4 J K^{-1} mol^{-1} for the adsorption of cellobiose on a hydrophilic polyacrylamide surface in an aqueous solution [46]. In this regard, another driving force caused by hydrophobic properties leads to the entropically favored adsorption, in addition to the formation of CH–π hydrogen bonds. This second factor can be ascribed to so-called hydrophobic interactions between the carbon and cellobiose. In previous works employing a combination of proteins/alkyl sepharose or cetyltrimethylammonium bromide/activated carbon, adsorption driven by hydrophobic interactions results in positive ΔS°_{ads} values [47, 48]. In an aqueous environment, the conformation and DF of water molecules surrounding apolar materials are restricted to maintain hydrogen-bonding networks (this structure is called iceberg model), and this state is entropically disfavored [49–51]. As a result, when multiple apolar materials exist in an aqueous solution, the water molecules are spontaneously unbound from the lipophilic surfaces to gain a higher DF, resulting in the association of apolar materials (Fig. 3.23). The surface of carbon is hydrophobic as indicated in cyclohexane adsorption and the axial plane of cellobiose also consists of lipophilic CH groups. Besides, cellobiose adsorbs on carbon surface via its hydrophobic plane through CH–π hydrogen bonds (vide supra). In a related work, a computation has revealed that a protein domain Pin WW interacts with sugar molecules through its aromatic residue by hydrophobic interactions in addition to the formation of CH–π hydrogen bonds [52]. To confirm the role of water solvent, DMSO was utilized as a solvent containing hydrophobic

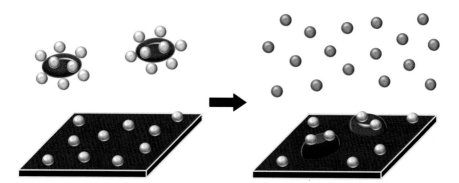

Fig. 3.23 Reorientation of water molecules associated with hydrophobic adsorption, so-called hydrophobic interaction. Legends: *black plates* hydrophobic material (in this study, carbon material); *red ellipsoids* hydrophobic solute (cellulosic molecules); *grey balls* water molecules restricted around hydrophobic surfaces; and *blue balls* free water molecules

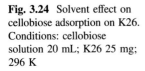

Fig. 3.24 Solvent effect on cellobiose adsorption on K26. Conditions: cellobiose solution 20 mL; K26 25 mg; 296 K

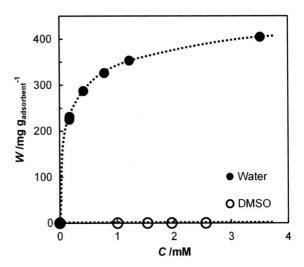

groups for the adsorption of cellobiose on K26, but cellobiose was not adsorbed on K26 at all (Fig. 3.24). There are two aspects why cellobiose does not adsorb on the carbon surface in DMSO. First, in contrast to water, DMSO medium does not cause hydrophobic interactions due to the absence of entropically disfavored iceberg structure; in other words, DMSO molecules surrounding cellobiose and carbon are not in an entropically unfavorable state. Second, DMSO has two hydrophobic methyl groups, which possibly form CH–π hydrogen bonds with the carbon surface, and competes with cellobiose in adsorption. Therefore, the author concludes that hydrophobic interactions are included in the adsorption of cellobiose on the carbon surface in water, which precedes the cellulose conversion. It is notable that the theoretical value of entropy change of cellobiose adsorption on carbon surface is calculated to be +12.1 ± 37 J K^{-1} mol^{-1} and roughly agrees with the experimental value (+23.5 ± 3.4 J K^{-1} mol^{-1}). The theoretical estimation of this adsorption entropy change value is summarized in Sect. 7.1.

The positive ΔS°_{ads} is profitable for adsorption of cellobiose on K26 in the temperature range shown in Table 3.7 (296–330 K), but also for that even at higher temperatures that are necessary for hydrolysis of cellulosic molecules. Specifically, based on Eq. 3.13, the ΔS°_{ads} value of +23.5 J K^{-1} mol^{-1} enhances the K_{ads} value by a factor of 16.9, which is independent of temperature, whereas the contribution of enthalpy to adsorption is gradually reduced with increasing temperature. The author demonstrated the adsorption of cellobiose on K26 even at 413 K; 230 mg of cellobiose adsorbed on 1 g of K26 from 1.7 mM cellobiose solution within 12 min. Note that cellulose was hydrolyzed into glucose by K26 over 24 h at near 413 K (see Sect. 2.3.3). Therefore, hydrophobic functionalities of both cellobiose and carbon materials play significant roles in the adsorption process at the wide range of temperatures.

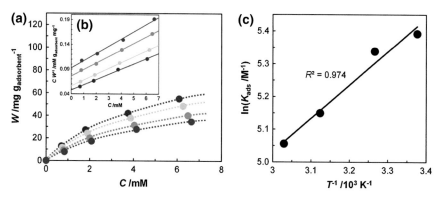

Fig. 3.25 Adsorption of glucose on K26: **a** isotherms, **b** Langmuir plots, and **c** a van't Hoff plot. Conditions: glucose aqueous solution 20 mL; K26 100 mg. Legends: *red* 296 K; *orange* 306 K; *green* 320 K; and *blue* 330 K

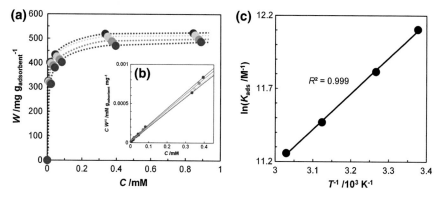

Fig. 3.26 Adsorption of cellotriose on K26: **a** isotherms, **b** Langmuir plots, and **c** a van't Hoff plot. Conditions: cellotriose aqueous solution 20 mL; K26 12.5 mg. Legends: *red* 296 K; *orange* 306 K; *green* 320 K; and *blue* 330 K

Other cellulosic molecules, glucose and cellotriose, were also tested for the adsorption on K26 at various temperatures to examine the effect of DP. Figures 3.25 and 3.26 depict the isotherms, Langmuir plots, and van't Hoff plots for glucose and cellotriose. The Langmuir parameters, K_{ads} and W_{max}, for both sugars are summarized in Table 3.7, which have been determined from the Langmuir plots (Figs. 3.25b and 3.26b) and Eq. 3.9. The W_{max} value at 296 K for glucose was 95.2 mg $g_{adsorbent}^{-1}$ (entry 36), corresponding to 171 m^2 of total adsorption cross-sectional area per 1 g of K26, and was the lowest among those for three glucans tested in this study [412 mg $g_{adsorbent}^{-1}$ (= 558 m^2, entry 40) for cellobiose and 527 mg $g_{adsorbent}^{-1}$ (= 735 m^2, entry 44) for cellotriose]. The significantly lower adsorption cross-sectional area of glucose implies that the adsorption sites for glucose might differ from those for oligomers. However, the glucose uptake for pristine K26 (12 μg $g_{adsorbent}^{-2}$ in 1.8 mM of glucose solution) was almost the same as that for K26-1273 (10 μg $g_{adsorbent}^{-2}$ at the same concentration); the adsorption of glucose onto the carbon surface is thus driven by hydrophobic functionalities as well as that of oligosaccharides. The K_{ads} values at 296 K drastically increased with increasing DP: 220 M^{-1} for glucose (entry 36); 5660 M^{-1} for cellobiose (entry 40); and 181000 M^{-1} for cellotriose (entry 44). This trend means that longer-chain oligosaccharides strongly interact with carbon materials. As well as cellobiose adsorption (entries 40–43), higher temperature made the adsorption of glucose and cellotriose more difficult (entries 36–39 and 44–47). The K_{ads} values at various temperatures led to the determination of $\Delta H°_{ads}$ and $\Delta S°_{ads}$ (Table 3.9). $\Delta H°_{ads}$ decreased in an almost linear fashion with increasing DP, reflecting the number of CH–π hydrogen bonds; each glucose unit has two or three CH groups in each axial plane. Meanwhile, $\Delta S°_{ads}$ gradually increased with the increase of DP. As mentioned previously, the positive $\Delta S°_{ads}$ is ascribed to the dynamical reorientation of water molecules surrounding sugars and K26, i.e., hydrophobic interactions. In hydrophobic interactions, $\Delta S°_{ads}$ becomes larger when larger numbers of water molecules were unbound from the apolar surfaces. In this regard, larger glucans enables the reorientation of greater numbers of water molecules. Hence, cellotriose affords the lowest $\Delta H°_{ads}$ and the highest $\Delta S°_{ads}$ among the three glucans employed in this study.

Finally, the affinity of cellulosic molecules with each other in water was examined to reveal whether multilayer adsorption of cellulosic molecules occurs on carbon surface. K26 and ball-milled cellulose were used as adsorbents for cellobiose at 296 K (Fig. 3.27a). Cellobiose strongly adsorbed on K26 as already discussed, whereas almost no adsorption was observed for ball-milled cellulose. This result is not attributed to a huge difference between specific surface areas of K26 and ball-milled cellulose, as the adsorption amounts of cellobiose standardized by specific surface area of ball-milled cellulose were also almost zero (Fig. 3.27b). The author additionally confirmed no adsorption of cellobiose on microcrystalline

Fig. 3.27 Adsorption isotherms for cellobiose on K26 (*closed circles*) and ball-milled cellulose (*open circles*) standardized by **a** mass of adsorbents and **b** specific surface area of adsorbents (2270 m^2 g^{-1} for K26 and 1.3 m^2 g^{-1} for ball-milled cellulose)

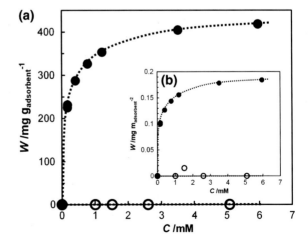

cellulose at all. The effect of DP on the adsorption was further investigated by testing cellohexaose as an adsorbate. The cellohexaose uptake was 333 mg $g_{adsorbent}^{-1}$ (= 0.147 mg $m_{adsorbent}^{-2}$) for K26 in a 0.035 mM cellohexaose aqueous solution, while the hexamer did not adsorb on ball-milled cellulose at all (<0.025 mg $m_{adsorbent}^{-2}$) under the same conditions. Therefore, the adsorption of cellulosic molecules on carbon surface is much more favorable than the association with each other. That is, multilayer adsorption of cellulosic molecules on adsorbents in the aqueous phase is negligible.

In brief summary of this section, the adsorption of cellulosic molecules on carbon surface is induced by hydrophobic planes of both adsorbate and adsorbent, namely the formation of CH–π hydrogen bonds as well as hydrophobic interactions. To the best of the author's knowledge, only direct interactions such as OH–O hydrogen bonds and CH–π ones have been considered so far [23, 24, 36, 37]. Here the author has revealed that entropically contribution derived from the rearrangement of water molecules surrounding hydrophobic materials is also included in the adsorption process in addition to enthalpically driven CH–π hydrogen bonds. The increase of DP enhances the affinity of cellulosic molecules with carbon. This fact enables to predict the role of carbon catalysts in cellulose hydrolysis: carbon attracts cellulose (long-chain polymer); the adsorbed cellulose undergoes hydrolysis; and produced shorter-chain oligosaccharides and glucose desorbs from carbon surface in the presence of long-chain oligomers. This is why K26 is significantly more active than model catalysts in the reaction (see Table 3.6). In other words, salicylic acid and phthalic acid do not have large aromatic planes, resulting in weaker affinity with substrates and lower catalytic activities than K26.

3.3.4 Proposed Mechanism of Cellulose Hydrolysis by Carbon Catalysts

From the results and discussion described above, the author proposes the reaction mechanism for cellulose hydrolysis by carbon catalysts as follows (Fig. 3.28). The first step (i) is the adsorption of cellulosic molecules onto the surface of carbon driven by both CH–π hydrogen bonds and hydrophobic interactions. This adsorption would help the subsequent interaction between hydrophilic functionalities. In the second step (ii), vicinal weak acids on carbon, the structure of which is similar to those of salicylic acid and phthalic acid, interact with the substrate through OH–O hydrogen bonds to activate a glycosidic bond. Then, in step (iii), the glycosidic bond is cleaved to form an intermediate, oxocarbenium species. In step (iv), a water molecule readily attacks the oxocarbenium species and a hydroxyl group is produced. Finally, in step (v), the divided cellulosic molecules desorb from the carbon surface due to weaker affinity of shorter-chain molecules with carbon surfaces.

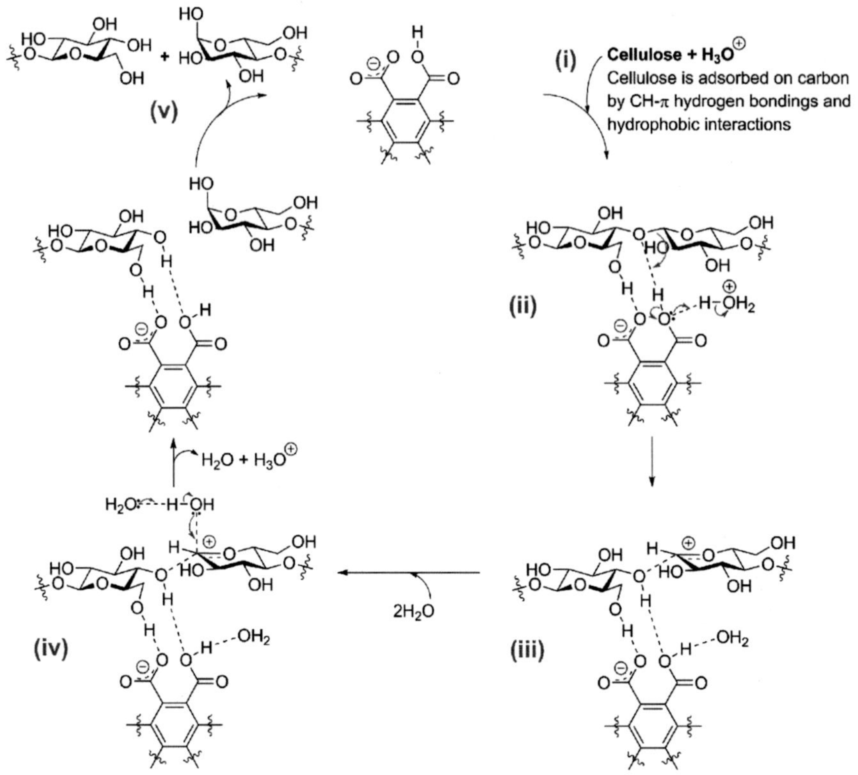

Fig. 3.28 Proposed reaction mechanism of cellulose hydrolysis by carbon catalysts

3.3.5 Hydrolysis of Cellulose by ZTC

The mechanistic study has indicated that both OFGs and aromatic rings of carbon materials play crucial roles in cellulose hydrolysis. Based on these insights, the author and collaborators have made efforts to prepare highly active carbon catalysts for cellulose hydrolysis.

Zeolite-templated carbon, denoted as ZTC (Fig. 3.29), consists of 3D-covalented warped nanographenes with a number of edges, thus possessing an ultra large surface area of ca. 4000 m^2 g^{-1} [9]. Since the nanographenes with huge surface area would be adsorption sites as discussed in Sect. 3.3.3, ZTC is expected to be a good adsorbent for cellulosic molecules. Fukuoka and Katz et al. has demonstrated that 800 mg of cellulosic molecules having 3600 g mol^{-1} of molecular mass adsorbed on 1 g of ZTC [38]. Besides, ZTC possibly contains large amounts of OFGs due to its unique structure containing many edges, onto which OFGs can be post-synthetically introduced. Indeed, Nishihara et al. has demonstrated the preparation of ZTC containing >10 wt% of O atoms as carboxylic acids and phenolic groups [9]. This report motivated the author and collaborators to utilize

Fig. 3.29 Structure of ZTC.
Reprinted from ref. [9],
Copyright 2009, with the
permission from Elsevier

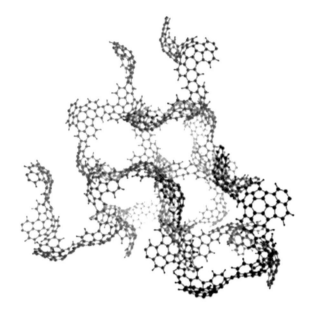

Table 3.10 Hydrolysis of individually ball-milled cellulose by ZTC

Entry	Catalyst	Conv./%	Yield based on carbon/%						
			Glucan		By-product				
			Glc[a]	Olg[b]	Frc[c]	Man[d]	Lev[e]	HMF[f]	Others[g]
32	None	28	4.6	15	0.5	0.6	0.2	1.8	5.3
54	ZTC	70	46	1.4	2.4	1.8	2.8	5.0	10
35	K26	60	36	2.5	2.7	2.6	2.1	3.4	11

Conditions individually ball-milled cellulose (0.9 L-pot) 324 mg; catalyst 50 mg; distilled water 40 mL; 503 K; rapid heating-cooling condition (Fig. 2.3)
[a]Glucose
[b]Water-soluble oligosaccharides (DP = mainly 2–6)
[c]Fructose
[d]Mannose
[e]Levoglucosan
[f]5-HMF
[g](Conversion) − (Total yield of the identified products)

electrolytically-oxidized ZTC as a catalyst in the hydrolysis of individually ball-milled cellulose (Table 3.10). The oxidized ZTC produced glucose in 46 % yield with 70 % conversion of cellulose (entry 54), and thus this material specifically showed a higher catalytic activity than other carbons tested in this work under the same conditions (see Table 2.1). This result clearly indicates that highly active carbon catalyst can be designed based on the mechanistic insights. Meanwhile, the oxidized ZTC material has given lower glucose yield (27 %) in the second run. Therefore, further investigations for oxidized ZTC including improvement of durability as well as its characterization are necessary in future work.

3.4 Conclusions

Adsorption and hydrolysis of cellulosic molecules over K26 have been investigated. Both hydrophobic and hydrophilic functionalities of carbon catalysts play significant roles for the reaction. The formation of CH–π hydrogen bonds and hydrophobic interactions lead the initial adsorption step of the substrate onto carbon surface. Then, vicinal weakly-acidic groups such as OH/COOH and COOH/COOH on carbon synergistically activate and hydrolyze the glycosidic bonds of cellulosic molecules. During the reaction, one functional group makes an OH–O hydrogen bond with a hydroxyl group of the substrate, and then another carboxylic acid hydrolyzes the glycosidic bonds. The good interactions between carbon catalysts and cellulosic molecules enable to efficiently hydrolyze glycosidic bonds even in the absence of strong acid sites.

References

1. Boehm H (1994) Some aspects of the surface chemistry of carbon blacks and other carbons. Carbon 32(5):759–769
2. Cardona-Martinez N, Dumesic JA (1992) Applications of adsorption microcalorimetry to the study of heterogeneous catalysis. Adv Catal 38:149–244
3. Lin H-P, Kao C-P, Mou C-Y, Liu S-B (2000) Counterion effect in acid synthesis of mesoporous silica materials. J Phys Chem B 104(33):7885–7894
4. Yang H, Zhang L, Zhong L, Yang Q, Li C (2007) Enhanced cooperative activation effect in the hydrolytic kinetic resolution of epoxides on [Co(salen)] catalysts confined in nanocages. Angew Chem Int Ed 46(36):6861–6865
5. Gazit OM, Charmot A, Katz A (2011) Grafted cellulose strands on the surface of silica: effect of environment on reactivity. Chem Commun 47(1):376–378
6. Gazit OM, Katz A (2011) Grafted Poly(1→4-β-glucan) strands on silica: a comparative study of surface reactivity as a function of grafting density. Langmuir 28(1):431–437
7. Gazit OM, Katz A (2013) Understanding the role of defect sites in glucan hydrolysis on surfaces. J Am Chem Soc 135(11):4398–4402
8. Loerbroks C, Rinaldi R, Thiel W (2013) The electronic nature of the 1,4-β-glycosidic bond and its chemical environment: DFT insights into cellulose chemistry. Chem Eur J 19 (48):16282–16294
9. Nishihara H, Yang Q-H, Hou P-X, Unno M, Yamauchi S, Saito R, Paredes JI, Martínez-Alonso A, Tascón JMD, Sato Y, Terauchi M, Kyotani T (2009) A possible buckybowl-like structure of zeolite templated carbon. Carbon 47(5):1220–1230
10. Figueiredo JL, Pereira MFR, Freitas MMA, Órfão J (1999) Modification of the surface chemistry of activated carbons. Carbon 37(9):1379–1389
11. Ishikawa S, Murayama T, Ohmura S, Sadakane M, Ueda W (2013) Synthesis of novel orthorhombic Mo and V based complex oxides coordinating alkylammonium cation in its heptagonal channel and their application as a catalyst. Chem Mater 25(11):2211–2219
12. Knight DS, White WB (1989) Characterization of diamond films by Raman spectroscopy. J Mater Res 4(02):385–393
13. Pimenta MA, Dresselhaus G, Dresselhaus MS, Cançado LG, Jorio A, Saito R (2007) Studying disorder in graphite-based systems by Raman spectroscopy. Phys Chem Chem Phys 9(11):1276–1290

14. Stephens PJ, Devlin FJ, Chabalowski CF, Frisch MJ (1994) Ab initio calculation of vibrational absorption and circular dichroism spectra using density functional force fields. J Phys Chem 98(45):11623–11627
15. Montgomery JA Jr, Frisch MJ, Ochterski JW, Petersson GA (1999) A complete basis set model chemistry. VI. Use of density functional geometries and frequencies. J Chem Phys 110(6):2822–2827
16. Tomasi J, Mennucci B, Cammi R (2005) Quantum mechanical continuum solvation models. Chem Rev 105(8):2999–3094
17. Becke AD (1993) Density-functional thermochemistry. III. The role of exact exchange. J Chem Phys 98(7):5648–5652
18. Lee C, Yang W, Parr RG (1988) Development of the Colle-Salvetti correlation-energy formula into a functional of the electron density. Phys Rev B 37(2):785–789
19. Hehre WJ, Ditchfield R, Pople JA (1972) Self-consistent molecular orbital methods. XII. Further extensions of gaussian-type basis sets for use in molecular orbital studies of organic molecules. J Chem Phys 56(5):2257–2261
20. Grimme S, Antony J, Ehrlich S, Krieg H (2010) A consistent and accurate *ab initio* parametrization of density functional dispersion correction (DFT-D) for the 94 elements H–Pu. J Chem Phys 132(15):154104
21. Jencks WP, Regenstein J (1968) Ionization constants of acids and bases. In: Sober MA (ed) Handbook of biochemistry. Chemical Rubber Company, Cleveland, pp 305–351
22. Capon B (1963) Intramolecular catalysis in glucoside hydrolysis. Tetrahedron Lett 4(14): 911–913
23. Suganuma S, Nakajima K, Kitano M, Yamaguchi D, Kato H, Hayashi S, Hara M (2008) Hydrolysis of cellulose by amorphous carbon bearing SO_3H, COOH, and OH groups. J Am Chem Soc 130(38):12787–12793
24. Kitano M, Yamaguchi D, Suganuma S, Nakajima K, Kato H, Hayashi S, Hara M (2009) Adsorption-enhanced hydrolysis of β-1,4-glucan on graphene-based amorphous carbon bearing SO_3H, COOH, and OH groups. Langmuir 25(9):5068–5075
25. Qrtiz P, Reguera E, Fernández-Bertrán J (2002) Study of the interaction of KF with carbohydrates in DMSO-d_6 by ^1H and ^{13}C NMR spectroscopy. J Fluorine Chem 113(1):7–12
26. Zhang J, Zhang H, Wu J, Zhang J, He J, Xiang J (2010) NMR spectroscopic studies of cellobiose solvation in EmimAc aimed to understand the dissolution mechanism of cellulose in ionic liquids. Phys Chem Chem Phys 12(8):1941–1947
27. Liepinsh E, Otting G, Wüthrich K (1992) NMR spectroscopy of hydroxyl protons in aqueous solutions of peptides and proteins. J Biomol NMR 2(5):447–465
28. Liepinsh E, Otting G (1996) Proton exchange rates from amino acid side chains–implications for image contrast. Magnet Reson Med 35(1):30–42
29. McCormick CL, Callais PA, Hutchinson BH Jr (1985) Solution studies of cellulose in lithium chloride and *N,N*-dimethylacetamide. Macromolecules 18(12):2394–2401
30. Röder T, Morgenstern B, Schelosky N, Glatter O (2001) Solutions of cellulose in *N,N*-dimethylacetamide/lithium chloride studied by light scattering methods. Polymer 42(16):6765–6773
31. Nevell TP, Upton WR (1976) The hydrolysis of cotton cellulose by hydrochloric acid in benzene. Carbohydr Res 49:163–174
32. Woods RJ, Andrews CW, Bowen JP (1992) Molecular mechanical investigations of the properties of oxocarbenium ions. 2. Application to glycoside hydrolysis. J Am Chem Soc 114(3):859–864
33. Charmot A, Katz A (2010) Unexpected phosphate salt-catalyzed hydrolysis of glycosidic bonds in model disaccharides: cellobiose and maltose. J Catal 276(1):1–5
34. Kögel-Knabner I (1997) ^{13}C and ^{15}N NMR spectroscopy as a tool in soil organic matter studies. Geoderma 80(3):243–270
35. Sing KSW (1985) Reporting physisorption data for gas/solid systems with special reference to the determination of surface area and porosity. Pure Appl Chem 57(4):603–619

36. Bui S, Verykios X, Mutharasan R (1985) In situ removal of ethanol from fermentation broths. 1. Selective adsorption characteristics. Ind Eng Chem Process Des Dev 24(4):1209–1213
37. Chung P-W, Charmot A, Gazit OM, Katz A (2012) Glucan adsorption on mesoporous carbon nanoparticles: effect of chain length and internal surface. Langmuir 28(43):15222–15232
38. Chung P-W, Yabushita M, To AT, Bae YJ, Jankolovits J, Kobayashi H, Fukuoka A, Katz A (2015) Long-chain glucan adsorption and depolymerization in zeolite-templated carbon catalysts. ACS Catal 5:6422–6425
39. Chung P-W, Charmot A, Click T, Lin Y, Bae YJ, Chu J-W, Katz A (2015) Importance of internal porosity for glucan adsorption in mesoporous carbon materials. Langmuir 31 (26):7288–7295
40. Tormo J, Lamed R, Chirino AJ, Morag E, Bayer EA, Shoham Y, Steitz TA (1996) Crystal structure of a bacterial family-III cellulose-binding domain: a general mechanism for attachment to cellulose. EMBO J 15(21):5739–5751
41. Igarashi K, Uchihashi T, Koivula A, Wada M, Kimura S, Okamoto T, Penttilä M, Ando T, Samejima M (2011) Traffic jams reduce hydrolytic efficiency of cellulase on cellulose surface. Science 333(6047):1279–1282
42. Zhu W, van de Graaf JM, van den Broeke LJP, Kapteijn F, Moulijn JA (1998) TEOM: a unique technique for measuring adsorption properties. Light alkanes in silicalite-1. Ind Eng Chem Res 37(5):1934–1942
43. Yong Z, Mata V, Rodrigues AE (2002) Adsorption of carbon dioxide at high temperature—a review. Sep Purif Technol 26(2):195–205
44. Sharma A, Kyotani T, Tomita A (2000) Comparison of structural parameters of PF carbon from XRD and HRTEM techniques. Carbon 38(14):1977–1984
45. Tsuzuki S, Honda K, Uchimaru T, Mikami M, Tanabe K (2000) The magnitude of the CH/π interaction between benzene and some model hydrocarbons. J Am Chem Soc 122(15):3746–3753
46. Brown W (1970) The separation of cellodextrins by gel permeation chromatography. J Chromatogr A 52:273–284
47. Gurses A, Yalcin M, Sozbilir M, Dogar C (2003) The investigation of adsorption thermodynamics and mechanism of a cationic surfactant, CTAB, onto powdered active carbon. Fuel Process Technol 81(1):57–66
48. Chen W-Y, Huang H-M, Lin C-C, Lin F-Y, Chan Y-C (2003) Effect of temperature on hydrophobic interaction between proteins and hydrophobic adsorbents: studies by isothermal titration calorimetry and the van't Hoff equation. Langmuir 19(22):9395–9403
49. Haselmeier R, Holz M, Marbach W, Weingaertner H (1995) Water dynamics near a dissolved noble gas. First direct experimental evidence for a retardation effect. J Phys Chem 99(8):2243–2246
50. Southall NT, Dill KA, Haymet ADJ (2002) A view of the hydrophobic effect. J Phys Chem B 106(3):521–533
51. Dill KA, Truskett TM, Vlachy V, Hribar-Lee B (2005) Modeling water, the hydrophobic effect, and ion solvation. Annu Rev Biophys Biomol Struct 34:173–199
52. Chen W, Enck S, Price JL, Powers DL, Powers ET, Wong C-H, Dyson HJ, Kelly JW (2013) Structural and energetic basis of carbohydrate-aromatic packing interactions in proteins. J Am Chem Soc 135(26):9877–9884

Chapter 4
Catalytic Depolymerization of Chitin to N-Acetylated Monomers

4.1 Introduction

Chitin (Fig. 4.1) is expected to be primary feedstock for organic nitrogen compounds, as chitin is the most abundant nitrogen-containing biomass consisting of a monomeric unit, N-acetylglucosamine (GlcNAc) [1]. GlcNAc can be produced by hydrolysis of chitin, and this compound is used as a biologically active agent, cosmetic and potential precursor to various kinds of organic nitrogen chemicals (Fig. 1.24) [2–7]. Methanolysis of chitin probably gives 1-O-methyl-N-acetylglucosamine (MeGlcNAc). MeGlcNAc inhibits hemagglutination, and thus this compound can suppress influenza and cancer [8, 9]. MeGlcNAc is also a promising precursor to biodegradable polyesters and polyamides [10], organocatalysts [11], ligands [12], and gelators [13]. Moreover, this N-acetylated compound is a useful molecule for further transformations in glycoscience since the hemiacetal group is protected by a methyl group [14]. It should be mentioned here that another type of monomer, a deacetylated compound glucosamine (GlcN), is also well known, but the removal of the acetyl group declines physiological activity and usability [15, 16]. Thus, the dissociation of glycosidic bonds of chitin with retention of acetyl group is favorable for efficient utilization of this abundant marine biomass resource.

To date, the selective synthesis of GlcNAc from chitin has been intensively studied; however, this reaction needs to overcome several obstacles as follows: chitin is highly recalcitrant; dissociation of glycosidic and amide bonds simultaneously takes place; and GlcNAc has a hemiacetal group that readily undergoes side reactions. The conventional methods using chitinase enzymes and concentrated HCl have overcome these issues. Chitinase enzymes depolymerize chitin to GlcNAc in a high yield of 77 % as they are excellently chemoselective and work under ambient conditions [17]. The drawback of this enzymatic hydrolysis is a long reaction time (10 days) to obtain such a high GlcNAc yield. Although 15–36 wt% concentration of HCl also can hydrolyze chitin with retaining N-acetyl group (ca. 65 % yield), this reaction system suffers from the serious corrosive property of

© Springer Science+Business Media Singapore 2016
M. Yabushita, *A Study on Catalytic Conversion of Non-Food Biomass into Chemicals*, Springer Theses, DOI 10.1007/978-981-10-0332-5_4

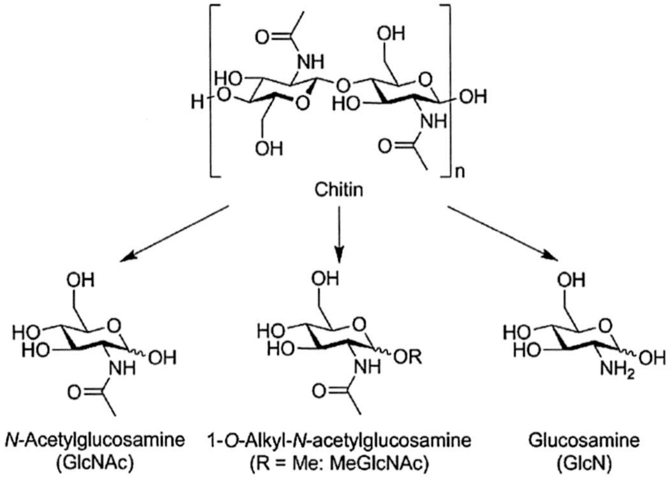

Fig. 4.1 Structure of chitin and possible depolymerization pathways

concentrated HCl and a huge amount of acidic waste [2]. Diluting HCl results in prominent side reactions including deacetylation in a longer reaction time due to the slower hydrolysis of glycosidic bonds [18]. These issues have hampered the utilization of chitin and monomers.

Hence, the selective depolymerization of chitin with preserving *N*-acetyl group has been a grand challenge in marine biomass refinery. Here the author shows a new strategy for the production of the *N*-acetylated monomers, GlcNAc and MeGlcNAc, from chitin without deacetylation by two-step depolymerization: (i) mechanical force-assisted hydrolysis and (ii) successive thermocatalytic solvolysis, hydrolysis and methanolysis in this study.

4.2 Experimental

4.2.1 Reagents

Chitin	First grade, Wako Pure Chemical Industries
D(+)-Cellobiose	Special grade, Kanto Chemical
Sulfuric acid	96–98 %, super special grade, Wako Pure Chemical Industries
Amberlyst 70	Sulfonic acid cation exchange resin, Organo
Nafion SAC-13	Silica-supported fluorinated sulfonic acid polymer, Sigma-Aldrich
JRC-Z-90H	Proton-type MFI zeolite, Si/Al ratio = 45, Catalysis Society of Japan, denoted as H-MFI

JRC-Z-HM90	Proton-type MOR zeolite, Si/Al = 45, Catalysis Society of Japan, denoted as H-MOR
H-BEA-50	Proton-type BEA zeolite, Si/Al = 25, Clariant Catalysts, denoted as H-BEA
K26	Alkali-activated carbon (not for sale), Showa Denko
N-Acetylglucosamine	First grade, Wako Pure Chemical Industries, denoted as GlcNAc
N, N'-Diacetylchitobiose	>98.0 %, Tokyo Chemical Industry
N, N', N''-Triacetylchitotriose	>98.0 %, Tokyo Chemical Industry
N, N', N'', N''', N''''-Pentaacetylchitopentaose	>97.0 %, Tokyo Chemical Industry
1-*O*-Methyl-α-*N*-acetylglucosamine	98 %, Toronto Research Chemicals, denoted as α-MeGlcNAc
1-*O*-Methyl-β-*N*-acetylglucosamine	98 %, Toronto Research Chemicals, denoted as β-MeGlcNAc
Acetic acid	>99.7 %, special grade, Wako Pure Chemical Industries
Methyl acetate	First grade, Wako Pure Chemical Industries
D(+)-Sorbitol	>97.0 %, Tokyo Chemical Industry
Distilled water	Wako Pure Chemical Industries
Methanol	Special grade, Wako Pure Chemical Industries
Hydrochloric acid methanolic solution	0.5 M, for volumetric analysis, Wako Pure Chemical Industries
Diethyl ether	First grade, Wako Pure Chemical Industries
Acetone	Special grade, Wako Pure Chemical Industries
Distilled water	For HPLC, Wako Pure Chemical Industries
Acetonitrile	For HPLC, Wako Pure Chemical Industries
Phosphoric acid	85 %, special grade, Kanto Chemical
Potassium dihydrogen phosphate solution	0.5 M, for HPLC, Wako Pure Chemical Industries
Deuterium oxide	For NMR, Acros Organics
Deuterated methanol	For NMR, Acros Organics
Maleic acid	>99 %, special grade, Wako Pure Chemical Industries
Potassium bromide	Crystal block, for IR, Wako Pure Chemical Industries
Helium gas	Alpha gas 2, Air Liquide Kogyo Gas
Hydrogen gas	>99.99 %, for gas chromatography (GC), generated from distilled water in an apparatus (GL Sciences, HG260)

4.2.2 Mechanocatalytic Hydrolysis of Chitin

Chitin (10 g) was dispersed in 25 mL of diethyl ether containing H_2SO_4 (0.59 g). After drying diethyl ether, 10 g of resulting powder was ball milled at 80 rpm for 96 h by using Al_2O_3 balls (ø1.5 cm, 2 kg) in a 3.6 L of ceramic pot. When using solid catalyst, pristine chitin 10 g and catalyst 1.23 g were milled together in the pot. For planetary ball-milling (Fritsch, Pulverisette 6), 4.9 g of H_2SO_4-impregnated chitin was treated at 500 rpm for 6 h in the presence of Al_2O_3 balls (ø5.0 mm, 100 g) in a 250 mL of Al_2O_3 pot, in which milling was stopped every 10 min for 10 min to decrease temperature in the pot. This sample prepared by planetary ball-milling is denoted as Chitin-H_2SO_4-PM hereafter.

The solubility of the chitin samples in water was determined as follows. The chitin sample (430 mg, containing 406 mg of chitin and 24 mg of H_2SO_4) was added into 40 mL of distilled water. After stirring and sonication for >5 min, the suspension was filtered with a PTFE membrane (0.1 μm mesh). The solid phase was dried in an oven at 383 K overnight and was weighed. The solubility was calculated from the difference between the amount of chitin used (= 406 mg) and that of dried residue. The water-soluble products were analyzed by LC/MS (Thermo Fischer Scientific, LCQ Fleet, APCI, the conditions were the same as those of HPLC described below). The structure of the products was also scrutinized by various NMR techniques (JEOL, JNM-ECX600, ^{1}H 600 MHz, ^{13}C 150 MHz): ^{1}H NMR spectroscopy; proton-decoupled ^{13}C NMR spectroscopy; distortionless enhancement by polarization transfer (DEPT); ^{13}C–^{1}H heteronuclear multiple quantum coherence (HMQC); and ^{13}C–^{1}H heteronuclear multiple bond correlation (HMBC). The LC/MS and NMR spectra are summarized in Sect. 7.2. Chitin samples themselves were also characterized by XRD (Rigaku, Ultima IV, Cu Kα radiation).

4.2.3 Thermocatalytic Solvolysis of Chitin Samples

Mix-milled chitin (456 mg, containing chitin 406 mg and catalyst 50 mg) or Chitin-H_2SO_4-PM (430 mg, containing 406 mg of chitin and 24 mg of H_2SO_4) and solvent (distilled water or methanol, 40 mL) were charged into a SUS316 high-pressure reactor (OM-Lab Tech, MMJ-100, 100 mL, its structure is the same as that of hastelloy C22 one depicted as Fig. 2.2). The temperature was raised to a certain temperature. After reaching the temperature, the reactor was rapidly cooled down to room temperature, namely rapid heating-cooling condition (Fig. 2.3). After the reaction, sorbitol (182 mg, 1 mmol) was added into the reaction solution as an internal standard. A portion of the reaction mixture (0.5 mL) was filtered by a Mini-UniPrep [Whatman, equipped with a PVDF membrane (0.2 μm mesh)], the amount of methyl acetate in the liquid phase was determined by GC (Shimadzu, GC-14B) with a ULBON HR-20 M capillary column (Shinwa Chemical Industries, ø0.25 mm × 25 m, film thickness: 0.25 μm). The liquid and solid phases of the remaining reaction mixture (*ca.* 39.5 mL) were separated by filtration using a PTFE membrane (0.1 µm mesh). The liquid phase was analyzed by the following procedure. After evaporating methanol, the reaction products were dissolved in 40 mL of distilled water. The products in the aqueous phase were quantified by HPLC [Shimadzu, LC10-AT*VP*, RI and UV (210 nm) detectors] with a SUGAR SH1011 column (Shodex, ø8 × 300 mm, mobile phase: water at 0.5 mL min^{-1}, 323 K), a Rezex RPM-Monosaccharide Pb++ column (Phenomenex, ø8 × 300 mm, mobile phase: water at 0.6 mL min^{-1}, 343 K), and a Asahipak NH2P-50 4E column [Shodex, ø4.6 × 250 mm, mobile phase: acetonitrile/water (7/3, vol/vol) at 0.5 mL min^{-1}, 303 K]. For the analysis of acetic acid by HPLC, a Synergi 4 µm Hydro-RP 80Å column [Phenomenex, ø4.6 × 250 mm, mobile phase: 40 mM potassium phosphate buffer solution (pH 2.9) at 0.8 mL min^{-1}, 303 K] was used. The author also conducted LC/MS (the conditions were the same as those of HPLC), IR [PerkinElmer, Spectrum 100, deuterated triglycine sulfate (DTGS) detector, transmission mode, KBr pellet], NMR (JEOL, JNM-ECX400, ^{1}H 400 MHz, ^{13}C 100 MHz, including ^{1}H NMR spectroscopy, ^{13}C NMR spectroscopy, DEPT, ^{13}C–^{1}H HMQC, and ^{13}C–^{1}H HMBC), and elemental analysis to identify reaction products. The LC/MS, IR, and NMR spectra of MeGlcNAc as well as their assignments are shown in Sect. 7.3.

4.2.4 Synthesis of MeGlcNAc

MeGlcNAc was synthesized from GlcNAc by Fischer glycosidation. GlcNAc (2.5 g) and HCl/methanol (50 mM, 40 mL) were charged into the hastelloy C22 high-pressure reactor (MMJ-100, OM-Lab Tech, 100 mL, Fig. 2.2). The temperature was raised to 423 K in 10 min and then rapidly decreased to room temperature. After removing HCl/methanol with a rotary evaporator, MeGlcNAc was recrystallized using methanol and acetonitrile as good and poor solvents, respectively. The resulting powder was filtered and washed with acetone repeatedly. After drying under vacuum overnight, 0.97 g of white MeGlcNAc powder with *ca.* 98 % of relative purity was obtained. The relative purity was determined by ^{1}H NMR (JEOL, JNM-ECX400, 400 MHz, repetition time: 7.7 s) using maleic acid as an internal standard.

4.3 Results and Discussion

4.3.1 Mechanocatalytic Hydrolysis of Chitin to Short-Chain Oligomers

The new strategy in this work is composed of two reactions, and the first step is mechanocatalytic hydrolysis of chitin to water-soluble short-chain oligomers (Fig. 4.2).

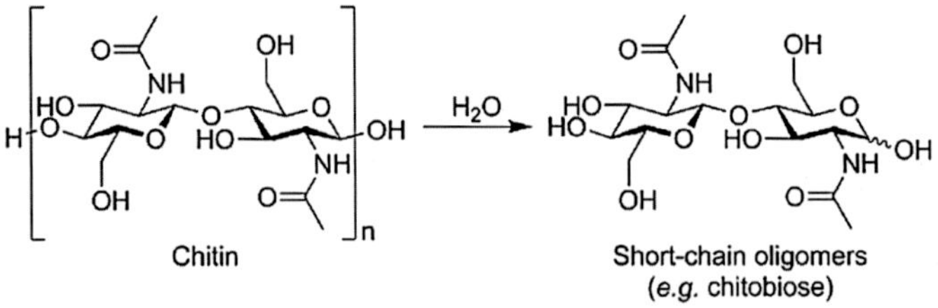

Fig. 4.2 Hydrolysis of chitin to water-soluble short-chain oligomers

In the reported works for polymer degradation, connecting points of polymers are activated and cleaved by mechanical force, as polymers can receive the macroscale power as tensile stress [19, 20]. Herein, the author has expected that mechanical force selectively divides glycosidic bonds connecting GlcNAc units rather than *N*-acetyl groups hanging from the units. Besides, mechanical reaction proceeds even under mild conditions [21–23], which is beneficial for suppression of side reactions.

At first, solid acids were tested for the mechanical conversion of chitin since heterogeneous catalysts are advantageous over homogenous ones in their ease of separation and reusability. Chitin and solid catalysts were ball-milled together at 80 rpm for 96 h (Table 4.1). In control experiments, neither pristine chitin nor ball-milled chitin was dissolved in water (entries 1 and 2). Amberlyst 70 gave *ca.* 88 % of soluble fractions, indicating the depolymerization of chitin to water-soluble compounds (entry 3). It is notable that chitin samples in this study contained *ca.* 1.5 wt% of physisorbed water, which was used for hydrolysis of chitin during milling process. Amberlyst 70 itself was unfortunately dissolved in water, similar to Fig. 2.9.

Table 4.1 Mechanocatalytic hydrolysis of chitin in the presence of acids

Entry	Catalyst	Treatment	Solubility/%	Yield/%	
				GlcNAc	Oligomers[a]
1	None	None	<0.1	<0.1	<0.1
2	None	Individual[b]	<0.1	<0.1	<0.1
3	Amberlyst 70	Mix[c]	54	5.4	31
4	Nafion SAC-13	Mix[c]	9.1	0.1	1.8
5	H-MFI	Mix[c]	10	0.2	1.7
6	H-MOR	Mix[c]	9.5	0.2	1.7
7	H-BEA	Mix[c]	9.5	0.2	1.9
8	K26	Mix[c]	7.8	0.1	1.4
9	H_2SO_4	Mix[c]	97	18	42[d]

Dissolution conditions chitin 406 mg; solid catalyst 50 mg or H_2SO_4 24 mg; distilled water 40 mL; 298 K
[a]Total yield of oligosaccharides with DP of 2–5
[b]Chitin was ball-milled without catalyst
[c]Chitin was ball-milled with catalyst
[d]Total yield of $(GlcNAc)_2$ and $(GlcNAc)_3$

Amberlyst 70 therefore does not have the advantage as a heterogeneous catalyst. The other solid acids, i.e., Nafion SAC-13, H-MFI, H-MOR, H-BEA, and K26, were almost inactive (entries 4–9). Hence, solid acids were not effective for the reaction. Next, H_2SO_4-impregnated chitin (S/C = 8.1) was ball-milled as a mechanocatalytic reaction (entry 9). 97 % of chitin was solubilized in water, indicating the formation of oligomers having low DP during the process; GlcNAc and oligomers were yielded in at least 60 %.

The conventional rolling pot-mill as described above required as long as 96 h of reaction time, which was a drawback in this system. The use of planetary ball-milling at 500 rpm shortened it to 6 h to solubilize chitin in >−99 % (Chitin-H_2SO_4-PM; Table 4.2, entry 13). In this case, the quantified products by HPLC were GlcNAc (4.7 % yield), $(GlcNAc)_2$ (7.8 %), $(GlcNAc)_3$ (11 %), $(GlcNAc)_4$ (9.7 %), and $(GlcNAc)_5$ (8.6 %), for which $(GlcNAc)_n$ represents an oligomer having n monomeric units. Additionally, the LC/MS and NMR analyses revealed the formation of longer and branched oligomers (see Sect. 7.2); the branched chains improve the solubility of oligomers [23]. These oligomers are useful as antitumor, immunoenhancing, and antimicrobial agents [24] as well as good precursors to monomers (see Sect. 4.3.2). Note that acetic acid was not contained in Chitin-H_2SO_4-PM, indicating that N-acetyl groups of chitin were completely preserved during the depolymerization. Accordingly, the ball-milling of chitin with H_2SO_4 selectively cleaved glycosidic bonds in chitin. Contrastively, the remarkable soluble fractions were not obtained in the other cases, i.e., pristine chitin (entry 1), H_2SO_4-impregnated chitin aged at 298 K for 6 h (entry 10, named Chitin-H_2SO_4), individually ball-milled chitin without H_2SO_4 (entry 11, Chitin-PM), and individually ball-milled chitin followed by the impregnation of H_2SO_4 and aging at 298 K for 6 h (entry 12, Chitin-PM-H_2SO_4). The planetary ball-milling treatment in the absence of H_2SO_4 did not cause hydrolysis but amorphization (Fig. 4.3) [25].

Table 4.2 Mechanocatalytic hydrolysis of chitin by H_2SO_4 and planetary ball-milling

Entry	Sample	Solubility/%	Yield/%	
			GlcNAc	Oligomers[a]
1	Chitin[b]	<0.1	<0.1	<0.1
10	Chitin-$H_2SO_4^c$	3.7	0.3	2.2
11	Chitin-PM[d]	5.2	0.1	0.8
12	Chitin-PM-$H_2SO_4^e$	7.8	0.2	3.0
13	Chitin-H_2SO_4-PM[f]	>99	4.7	37

Dissolution conditions chitin 406 mg; H_2SO_4 24 mg; distilled water 40 mL; 298 K
[a]Total yield of oligosaccharides with DP of 2–5
[b]Pristine chitin
[c]H_2SO_4 was impregnated on chitin, and then the sample was aged at 298 K for 6 h
[d]Chitin was planetary ball-milled without H_2SO_4
[e]Chitin was planetary ball-milled without H_2SO_4, followed by the impregnation of H_2SO_4 and aging at 298 K for 6 h
[f]H_2SO_4-impregnated chitin was planetary ball-milled

Fig. 4.3 XRD patterns of chitin samples. The sharp peak at 19.8° is derived from the crystalline structure of chitin [25]. After planetary ball-milling with/without H₂SO₄, amorphous chitin was formed

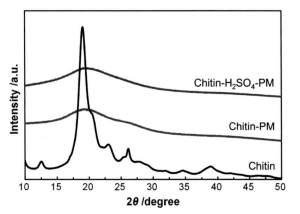

Both mechanical force and H_2SO_4 catalyst are necessary for the depolymerization of chitin, indicating mechanocatalytic reaction.

Figure 4.4 depicts the proposed schematic of the mechanocatalytic reaction. The mechanical force collapses the packing structure of chitin (i.e., amorphization), enhancing the accessibility of H_2SO_4 to reactive sites in chitin such as glycosidic bonds and amide bonds. The long chain of chitin (glycosidic bonds) receives macro-power as tensile stress to undergo activation and hydrolysis, standing in stark contrast to the short chains (*N*-acetyl groups).

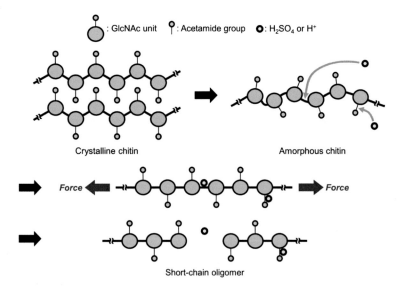

Fig. 4.4 Proposed schematic of mechanocatalytic hydrolysis of chitin

4.3.2 Thermocatalytic Solvolysis of Oligomers to N-Acetylated Monomers

The second step is thermocatalytic solvolysis, hydrolysis, and methanolysis in this study, to produce the respective N-acetylated monomers from the oligomers produced in the first step (Fig. 4.5).

The hydrolysis of the oligomers in Chitin-H_2SO_4-PM was conducted in water. The catalyst in this reaction was H_2SO_4 (S/C = 8.1) that was impregnated before the mechanocatalytic reaction. The hydrolysis at 463 K provided 32 % yield of GlcNAc (Table 4.3, entry 14). The larger amount of H_2SO_4 (S/C = 2.0) and lower reaction temperature (443 K) improved the yield of GlcNAc up to 53 % (entry 15).

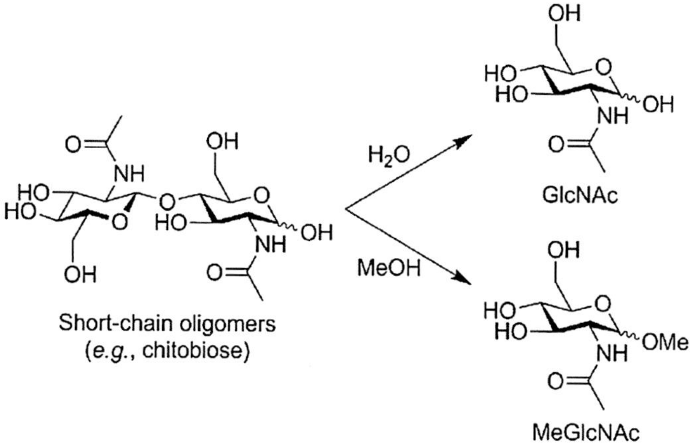

Fig. 4.5 Solvolysis of short-chain oligomers to N-acetylated monomers

Table 4.3 Thermocatalytic solvolysis of chitin samples to N-acetylated monomers

Entry	Sample	S/C	Temp./K	Solvent	Yield/%		
					GlcNAc	MeGlcNAc	Others[a]
14	Chitin-H_2SO_4-PM	8.1	463	Water	32	–	68
15	Chitin-H_2SO_4-PM	2.0	443	Water	53	–	47
16	Chitin	–[b]	463	Methanol	–	<0.1	<0.1
17	Chitin-H_2SO_4	8.1	463	Methanol	–	4.0	1.7
18	Chitin-PM	–[b]	463	Methanol	–	<0.1	2.2
19	Chitin-PM-H_2SO_4	8.1	463	Methanol	–	16	4.6
20	Chitin-H_2SO_4-PM	8.1	463	Methanol	–	68	27
21	Chitin-H_2SO_4-PM	4.1	463	Methanol	–	70	28

Conditions chitin 406 mg; H_2SO_4 24 mg; solvent 40 mL; rapid heating-cooling condition (see Fig. 2.3)
[a]Total yield of soluble products other than the N-acetylated monomers
[b]The reaction was conducted without H_2SO_4

However, by-products were notably formed in both cases. This is possibly due to instability of a hemiacetal group in GlcNAc under hydrothermal conditions. In a stability test, only 17 % of GlcNAc survived at 463 K (Table 4.4, entry 22). It is noteworthy that acetic acid was not observed (<1 %) in this stability test, indicating that the hemiacetal group readily underwent side reactions.

In contrast to GlcNAc, MeGlcNAc was thermally stable since 94 % of MeGlcNAc was recovered after the exposure to the reaction conditions (Table 4.4, entry 23). Regarding the structure of MeGlcNAc, hemiacetal group is protected by a methyl group, which has successfully suppressed side reactions. In a related work, Yan et al. operated the alcoholysis of chitin in ethylene glycol in the presence of H_2SO_4 (S/C = 0.9); however, the major product was a deacetylated monomer, 1-*O*-(2-hydroxyethyl)-2-amino-2-deoxyglucopyranoside (HADP, Fig. 1.25a), due to harsh conditions to depolymerize robust chitin [26]. Herein, Chitin-H_2SO_4-PM was subjected to methanolysis to obtain chemically stable MeGlcNAc in a good yield; MeGlcNAc was yielded in 68 % (S/C = 8.1, entry 20). The produced MeGlcNAc was identified by combining several techniques including NMR, IR, LC/MS, HPLC, and elemental analysis (see Sect. 7.3). The ratio of α- to β-MeGlcNAc was 5.0 determined from ^{1}H NMR, and this good stereoselectivity is favorable for further applications; the origin of this ratio is discussed below. Turnover number (TON) for the production of MeGlcNAc was 5.6, showing the *catalytic* reaction. This fashion is completely different from the reported work requiring massive amount of acids to hydrolyze chitin, resulting in *non-catalytic* reactions (TON < 0. 01) [2]. Only 0.90 % of methyl acetate was observed by GC, and thus most of *N*-acetyl groups in the sample were maintained during the methanolysis. The lower S/C ratio increased the yield of MeGlcNAc up to 70 % (S/C = 4.0, entry 21). In control experiments employing the other chitin samples (pristine chitin, Chitin-H_2SO_4, Chitin-PM, and Chitin-PM-H_2SO_4), the yields of MeGlcNAc were at most 16 % (entries 16–19). These results clearly show that the mechanocatalytic depolymerization of chitin antecedent to thermocatalytic methanolysis is essential for the high-yielding synthesis of MeGlcNAc. It is noteworthy that the author further conducted the methanolysis of ball-milled chitin containing solid catalysts that were used in the water-solubility tests (see Sect. 4.3.1), but all samples did not produce MeGlcNAc in remarkable yields.

The reaction mechanism explains the stereoselective production of α-MeGlcNAc (α/β = 5.0). The solvolysis of glycosidic bonds proceeds through S_N1 mechanism via the formation of oxocarbenium intermediates as discussed in Sect. 3.3.2. The oxocarbenium species could undergo both front- and backside attacks by a

Table 4.4 Stability test of *N*-acetylated monomers

Entry	Substrate	Recovery/%
22[a]	GlcNAc	17
23[b]	MeGlcNAc	94

[a]*Conditions* GlcNAc 442 mg; distilled water 40 mL; 463 K; rapid heating-cooling condition (see Fig. 2.3)
[b]*Conditions* MeGlcNAc 470 mg; methanol 40 mL; 463 K; rapid heating-cooling condition (see Fig. 2.3)

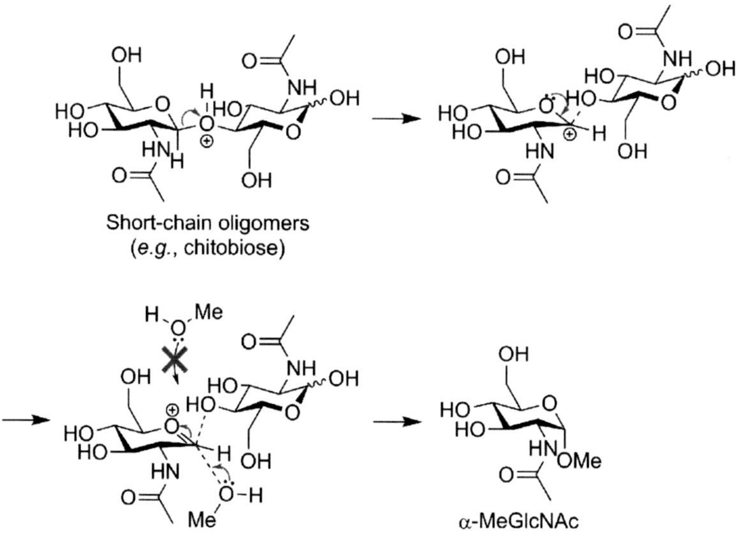

Fig. 4.6 Proposed reaction mechanism of methanolysis to produce α-MeGlcNAc via oxocarbenium intermediates. Due to the presence of leaving group, the nucleophilic attack from the upper side in this figure is limited

nucleophile (i.e., methanol), giving a mixture of α- and β-MeGlcNAc with the ratio of 1.0. Nonetheless, α-MeGlcNAc was predominantly formed, showing an inversion of stereochemistry of chitin (= β-1,4-glycosidic bond). This result implies that methanol probably attacks oxocarbenium species from backside before the leaving group (counterpart of chitin oligomer) diffuses outside of solvent cage due to very high concentration of methanol (18 M at 463 K [27], Fig. 4.6) [28, 29]. Although an anomeric effect is also considered, this is not the main reason since such inversions of stereochemistry has been observed in similar methanolysis reactions (see Table 3.5).

4.4 Conclusions

A two-step depolymerization of chitin has been operated to synthesize N-acetylated monomers, GlcNAc, and MeGlcNAc. In the first mechanocatalytic hydrolysis step, chitin is transformed into water-soluble oligomers with low DP for 6 h. The following thermocatalytic solvolysis produces GlcNAc and MeGlcNAc in good yields within 1 h. These reactions require trace amount of H_2SO_4 (0.075 %), which is by 99.8 % less than those used in conventional methods [2]. Another advantage of this two-step depolymerization is a shorter reaction time (7 h), compared to enzymatic reactions (10 days) [17]. Hence, this new method has overcome the issues in conventional systems. The author believes that the two-step depolymerization system has the potential for the efficient utilization of not only chitin, but also various polymers.

References

1. Chen X, Yan N (2014) Novel catalytic systems to convert chitin and lignin into valuable chemicals. Catal Surv Asia 18(4):164–176
2. Chen J-K, Shen C-R, Liu C-L (2010) *N*-Acetylglucosamine: production and applications. Mar Drugs 8(9):2493–2516
3. Omari KW, Dodot L, Kerton FM (2012) A simple one-pot dehydration process to convert *N*-acetyl-D-glucosamine into a nitrogen-containing compound, 3-acetamido-5-acetylfuran. ChemSusChem 5(9):1767–1772
4. Wang Y, Pedersen CM, Deng T, Qiao Y, Hou X (2013) Direct conversion of chitin biomass to 5-hydroxymethylfurfural in concentrated $ZnCl_2$ aqueous solution. Bioresour Technol 143:384–390
5. Osada M, Kikuta K, Yoshida K, Totani K, Ogata M, Usui T (2013) Non-catalytic synthesis of Chromogen I and III from *N*-acetyl-D-glucosamine in high-temperature water. Green Chem 15 (10):2960–2966
6. Ohmi Y, Nishimura S, Ebitani K (2013) Synthesis of α-amino acids from glucosamine-HCl and its derivatives by aerobic oxidation in water catalyzed by Au nanoparticles on basic supports. ChemSusChem 6(12):2259–2262
7. Chen X, Chew SL, Kerton FM, Yan N (2014) Direct conversion of chitin into a *N*-containing furan derivative. Green Chem 16(4):2204–2212
8. Tabary F, Font J, Bourrillon R (1987) Isolation, molecular and biological properties of a lectin from rice embryo: relationship with wheat germ agglutinin properties. Arch Biochem Biophys 259(1):79–88
9. Kochibe N, Matta KL (1989) Purification and properties of an *N*-acetylglucosamine-specific lectin from *Psathyrella velutina* mushroom. J Biol Chem 264(1):173–177
10. Fujii S, Kondo Y, Matsui M, Ichihashi K (1993) Methods for the production of sugar-based synthetic polymers (Japanese title: Toushitsu gousei koubunshi oyobi sono seizouhou). JP Patent 5-178904
11. Agarwal J, Peddinti RK (2012) Synthesis and characterization of monosaccharide derivatives and application of sugar-based prolinamides in asymmetric synthesis. Eur J Org Chem 32:6390–6406
12. RajanBabu TV, Ayers TA, Halliday GA, You KK, Calabrese JC (1997) Carbohydrate phosphinites as practical ligands in asymmetric catalysis: electronic effects and dependence of backbone chirality in Rh-catalyzed asymmetric hydrogenations. Synthesis of *R*- or *S*-amino acids using natural sugars as ligand precursors. J Org Chem 62(17):6012–6028
13. Goyal N, Cheuk S, Wang G (2010) Synthesis and characterization of D-glucosamine-derived low molecular weight gelators. Tetrahedron 66(32):5962–5971
14. Koeller KM, Wong C-H (2000) Emerging themes in medicinal glycoscience. Nat Biotechnol 18(8):835–841
15. Shikhman AR, Kuhn K, Alaaeddine N, Lotz M (2001) *N*-Acetylglucosamine prevents IL-1β-mediated activation of human chondrocytes. J Immunol 166(8):5155–5160
16. Álvarez-Añorve LI, Calcagno ML, Plumbridge J (2005) Why does *Escherichia coli* grow more slowly on glucosamine than on *N*-acetylglucosamine? Effects of enzyme levels and allosteric activation of GlcN6P deaminase (NagB) on growth rates. J Bacteriol 187(9): 2974–2982

17. Sashiwa H, Fujishima S, Yamano N, Kawasaki N, Nakayama A, Muraki E, Hiraga K, Oda K, Aiba S (2002) Production of N-acetyl-D-glucosamine from α-chitin by crude enzymes from *Aeromonas hydrophila* H-2330. Carbohydr Res 337(8):761–763
18. Einbu A, Vårum KM (2007) Depolymerization and de-N-acetylation of chitin oligomers in hydrochloric acid. Biomacromolecules 8(1):309–314
19. Beyer MK, Clausen-Schaumann H (2005) Mechanochemistry: the mechanical activation of covalent bonds. Chem Rev 105(8):2921–2948
20. Davis DA, Hamilton A, Yang J, Cremar LD, Van Gough D, Potisek SL, Ong MT, Braun PV, Martínez TJ, White SR, Moore JS, Sottos NR (2009) Force-induced activation of covalent bonds in mechanoresponsive polymeric materials. Nature 459(7243):68–72
21. Hick SM, Griebel C, Restrepo DT, Truitt JH, Buker EJ, Bylda C, Blair RG (2010) Mechanocatalysis for biomass-derived chemicals and fuels. Green Chem 12(3):468–474
22. Meine N, Rinaldi R, Schüth F (2012) Solvent-free catalytic depolymerization of cellulose to water-soluble oligosaccharides. ChemSusChem 5(8):1449–1454
23. Shrotri A, Lambert LK, Tanksale A, Beltramini J (2013) Mechanical depolymerisation of acidulated cellulose: understanding the solubility of high molecular weight oligomers. Green Chem 15(10):2761–2768
24. Shahidi F, Arachchi JKV, Jeon Y-J (1999) Food applications of chitin and chitosans. Trends Food Sci Technol 10(2):37–51
25. Osada M, Miura C, Nakagawa YS, Kaihara M, Nikaido M, Totani K (2013) Effects of supercritical water and mechanochemical grinding treatments on physicochemical properties of chitin. Carbohydr Polym 92(2):1573–1578
26. Pierson Y, Chen X, Bobbink FD, Zhang J, Yan N (2014) Acid-catalyzed chitin liquefaction in ethylene glycol. ACS Sustainable Chem Eng 2(8):2081–2089
27. Goodwin RD (1987) Methanol thermodynamic properties from 176 to 673 K at pressures to 700 Bar. J Phys Chem Ref Data 16(4):799–892
28. Winstein S, Clippinger E, Fainberg AH, Heck R, Robinson GC (1956) Salt effects and ion pairs in solvolysis and related reactions. III. Common ion rate depression and exchange of anions during acetolysis. J Am Chem Soc 78(2):328–335
29. Chan J, Tang A, Bennet AJ (2012) A stepwise solvent-promoted S_Ni reaction of α-D-glucopyranosyl fluoride: mechanistic implications for retaining glycosyltransferases. J Am Chem Soc 134(2):1212–1220

Chapter 5
Acid-Catalyzed Dehydration of Sorbitol to 1,4-Sorbitan

5.1 Introduction

In Chap. 2, the high-yielding one-pot synthesis of glucose from cellulose has been achieved using carbon catalysts and mix-milling pretreatment. In this chapter, the author deals with reaction routes from glucose to useful chemicals in order to emphasize the achievement of this thesis work. Glucose is transformed into a variety of valuable compounds via sorbitol, e.g., medicines, plastics, surfactants, and fuels (Figs. 1.7 and 1.10) [1–3]. Particularly, dehydration of sorbitol provides 1,4-sorbitan (1,4-anhydrosorbitol, Fig. 5.1). 1,4-Sorbitan is a precursor to environmentally benign surfactants [4], which are also used as emulsifying agents in food and pharmaceutical industries [5]. The annual demand of these surfactants is more than 10,000 tons worldwide [2].

1,4-Sorbitan is produced by the dehydration of sorbitol using acid catalysts or hot-compressed water, in which 1,4-sorbitan readily undergoes further dehydration to give isosorbide (1,4:3,6-dianhydrosorbitol) [6–20]. This hampers the selective and high-yielding synthesis of 1,4-sorbitan from sorbitol. Currently, H_2SO_4 is employed as a catalyst for sorbitol dehydration in industrial processes, since this low price acid gives 1,4-sorbitan in a relatively high yield (58 %) [6, 7, 10]. Although it is preferable to use heterogeneous catalysts owing to their ease of separation and reuse as well as no corrosivity, further improvement is necessary for replacing H_2SO_4. Herein, the first purpose of this chapter is selective production of 1,4-sorbitan from sorbitol over solid acids. The second objective is revealing origins of high selectivity of catalysts through mechanistic study since such insights are useful and helpful to design more active catalysts.

© Springer Science+Business Media Singapore 2016
M. Yabushita, *A Study on Catalytic Conversion of Non-Food Biomass into Chemicals*, Springer Theses, DOI 10.1007/978-981-10-0332-5_5

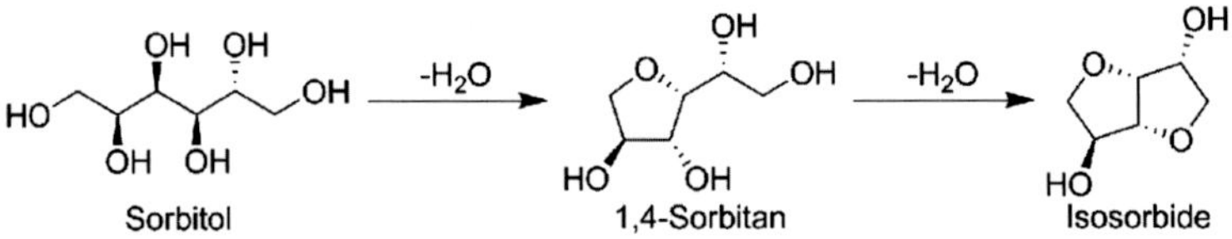

Fig. 5.1 Dehydration of sorbitol

5.2 Experimental

5.2.1 Reagents

D(+)-Sorbitol	>97.0 %, Tokyo Chemical Industry
Sulfuric acid	96–98 %, super special grade, Wako Pure Chemical Industries
JRC-ZRO-2	Zirconium(IV) hydroxide, $ZrO_{1.1}(OH)_{1.8}$, Catalysis Society of Japan
Ammonium sulfate	Special grade, Wako Pure Chemical Industries
Amberlyst 70	Sulfonic acid cation exchange resin, Organo
Nafion SAC-13	Silica-supported fluorinated sulfonic acid polymer, Sigma-Aldrich
CBV 780	Proton-type FAU zeolite, Si/Al = 40, Zeolyst, denoted as H-FAU
JRC-Z5-90H	Proton-type MFI zeolite, Si/Al = 45, Catalysis Society of Japan, denoted as H-MFI
JRC-Z-HM90	Proton-type MOR zeolite, Si/Al = 45, Catalysis Society of Japan, denoted as H-MOR
Silica-alumina	Grade 135, Sigma-Aldrich, denoted as SiO_2-Al_2O_3
1,4-Anhydrosorbitol	97 %, Toronto Research Chemicals, denoted as 1,4-sorbitan
D-1,4:3,6-Dianhydrosorbitol	>98.0 %, Tokyo Chemical Industry, denoted as isosorbide
1,5-Anhydrosorbitol	Toronto Research Chemicals, denoted as 1,5-sorbitan
1,5-Anhydromannitol	Toronto Research Chemicals, denoted as 1,5-mannitan
2,5-Anhydromannitol	Sigma-Aldrich, denoted as 2,5-mannitan
Distilled water	Wako Pure Chemical Industries
Diethyl ether	First grade, Wako Pure Chemical Industries
Dimethyl sulfoxide	Special grade, Wako Pure Chemical Industries, denoted as DMSO
Distilled water	For HPLC, Wako Pure Chemical Industries
Sodium hydrogen carbonate	Special grade, Wako Pure Chemical Industries
Sodium carbonate	Special grade, Wako Pure Chemical Industries

Deuterium oxide		For NMR, Acros Organics
Deuterated sulfoxide	dimethyl	For NMR, Acros Organics, denoted as DMSO-d$_6$
Deuterated methanol		For NMR, Acros Organics

5.2.2 Preparation of Sulfated Zirconia

Sulfated zirconia, denoted as SZ, was prepared as follows [21]. 0.5 g of $ZrO_{1.1}(OH)_{1.8}$ was dispersed in 10 mL of distilled water in an eggplant flask (100 mL). Then, a 0.52 M $(NH_4)_2SO_4$ aqueous solution (10 mL, corresponding to 18 wt% S in SZ) was added dropwise to the suspension over ca. 2 min with stirring by hand in an ultrasonic generator. After drying the mixture under vacuum at room temperature for ≥18 h and crashing on an Al_2O_3 mortar, the resulting white solid was heated to 773 K in 6 h and calcined at the temperature for 4 h under air in an electric furnace (Denken-Highdental, KDF-S90). Finally, 0.60 g of white powder (SZ) was obtained. Similarly, ZrO_2 was prepared by the calcination of $ZrO_{1.1}(OH)_{1.8}$ under the above conditions.

The content of sulfur in SZ was determined to be 20 wt% by energy dispersive X-ray spectroscopy (EDX, Shimadzu, EDX-720) and thermal analysis [22], which was consistent with the initial amount of sulfur in the catalyst preparation (18 wt%).

5.2.3 Dehydration of Sorbitol

Dehydration reaction was conducted without solvent in an eggplant flask (50 mL). Sorbitol 182 mg (1 mmol) and solid acid 50 mg were charged into the flask, and thus S/C was 3.6. The flask was immersed in an oil bath at 403 K; the actual temperature of reaction mixture was ca. 400 K [20]. After the reaction, 20 mL of water was added to the mixture at room temperature to extract products, and the water-soluble compounds were analyzed by HPLC [Shimadzu, LC10-AT*VP*, RI and UV (210 nm) detectors] with a SUGAR SH1011 column (Shodex, ø8 × 300 mm, mobile phase: water at 0.5 mL min^{-1}, 323 K) and a Rezex RPM-Monosaccharide Pb++ column (Phenomenex, ø7.8 × 300 mm, mobile phase: water at 0.6 mL min^{-1}, 343 K). Yields of products and amount of unreacted sorbitol were determined by an absolute calibration method.

The reuse experiment for SZ was conducted by the following procedure. The dehydration reaction was performed in the same manner as described above. After the reaction, 20 mL of DMSO was added to extract reaction products instead of distilled water, and then the suspension was filtered with a PTFE membrane (0.1 μm mesh). The residue was repeatedly washed with diethyl ether and then dried in an oven at 383 K. The resulting powder was used for the next reaction, in which S/C was kept at 3.6. Note that water is not a useful extractant for this

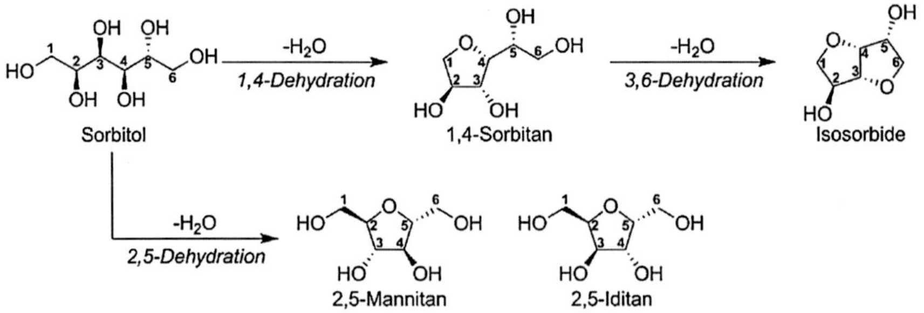

Fig. 5.2 Structures of identified products

experiment since *liquid water* readily causes leaching of acid species on SZ, unlike the *water vapor* formed during the dehydration of sorbitol.

The amount of SO_4^{2-} leached from SZ was determined by ion chromatography. Soluble components were extracted with DMSO after the reaction. The DMSO solution was diluted 2–3 times by distilled water and was analyzed by ion chromatography (Dionex, ICS-3000, conductivity detector) with an IonPac AS12A column [Dionex, ø4 × 200 mm, mobile phase: $NaHCO_3$ (2.7 mM)/Na_2CO_3 (0.3 mM) buffer at 1.5 mL min^{-1}, 308 K].

For dehydration of sorbitol by H_2SO_4, H_2SO_4 was impregnated on sorbitol before the reaction. A certain amount of H_2SO_4 was dissolved in 10 mL of diethyl ether, and then sorbitol (3.64 g) was dispersed in the solution. By removing diethyl ether under vacuum at room temperature, H_2SO_4-impregnated sorbitol was obtained. In the dehydration reaction, the H_2SO_4-impregnated sorbitol (containing sorbitol 1 mmol and H_2SO_4 1.0 or 6.5 μmol) was charged into the flask.

5.2.4   Product Identification

The reaction products were identified by means of NMR [JEOL, JNM-ECX400 ([1]H 400 MHz, [13]C 100 MHz) and JNM-ECX600 ([1]H 600 MHz, [13]C 150 MHz)] with various measurement methods: ^{1}H NMR, proton-decoupled ^{13}C NMR, DEPT, [13]C–[1]H HMQC. LC/MS (Thermo Fisher Scientific, LCQ Fleet, APCI) was also operated under the same conditions as those of HPLC shown above. The structures of the identified products [i.e., 1,4-sorbitan, isosorbide, 2,5-mannitan, and 2,5-iditan (2,5-anhydroiditol)] are represented in Fig. 5.2. Mono-anhydrohexitols, denoted as *AH1* and *AH2*, were also observed by HPLC. Although 2,5-iditan and *AH1* were not separated by HPLC, they were distinguished by NMR spectroscopy. The NMR spectra, LC/MS spectra, HPLC charts, and assignments are summarized in Sect. 7.4

.

5.3 Results and Discussion

5.3.1 Dehydration of Sorbitol Catalyzed by Acids

A variety of acids were tested for dehydration of sorbitol at 403 K for 120 min without solvent (Table 5.1). Relatively low melting points of sorbitol, 1,4-sorbitan, and isosorbide (368, 385, and 336 K, respectively) enabled to operate the dehydration under neat conditions. Sorbitol did not undergo any reactions in the absence of catalysts even for a longer reaction time of 180 min (entry 1). Sorbitol is stable and requires activation by catalysts for its transformation. SZ afforded 96 % conversion of sorbitol, and the main product was 1,4-sorbitan with 60 % yield and 62 % selectivity based on the conversion of sorbitol (entry 2). The successive dehydration to isosorbide also took place (11 % yield). Other products were 2,5-mannitan (3.9 %, Fig. 5.2), 2,5-iditan, $AH1$, and $AH2$ (1.1 %). The total yield of 2,5-iditan and $AH1$ was 8.9 %.

In Table 5.1, Amberlyst 70 and Nafion SAC-13 converted almost all sorbitol, but the main product was isosorbide due to the subsequent dehydration of 1,4-sorbitan (entries 5 and 6). Over Nafion SAC-13, the yield of 1,4-sorbitan was at most 38 % even under the optimized reaction time at 403 K (Fig. 5.3), and it was clearly lower than that by SZ. The colors of these catalysts turned dark brown due to coking during the reaction. Zeolite catalysts (H-FAU, H-MFI, and H-MOR) also gave low

Table 5.1 Dehydration of sorbitol by a variety of acids

Entry	Catalyst	Time/min	Conv./%	Yield/%				
				1,4-Srb[a]	2,5-Idt + AH1[b]	Iso[c]	2,5-Man[d]	AH2
1	None	180	<1	0.0	0.0	0.0	0.0	0.0
2	SZ	120	96	60	8.9	11	3.9	1.1
3	SZ	120	64	40	7.4	3.8	2.9	2.8
4	ZrO$_2$	120	2.3	0.0	0.0	0.0	0.0	0.0
5	Amberlyst 70	120	99	1.7	0.8	63	1.6	0.2
6	Nafion SAC-13	120	>99	9.0	9.1	66	4.2	0.2
7	H-FAU	120	76	31	11	14	7.2	2.0
8	H-MFI	120	55	8.9	2.2	30	1.1	4.2
9	H-MOR	120	24	12	3.0	2.8	1.0	1.2
10	SiO$_2$-Al$_2$O$_3$	120	2.4	0.6	0.0	0.0	0.0	0.0
11	H$_2$SO$_4$	120	77	56	7.5	6.1	2.8	2.9

Conditions sorbitol 182 mg (1 mmol); solid acid 50 mg or H$_2$SO$_4$ 1.0 μmol; 403 K
[a]1,4-Sorbitan
[b]Total yield of 2,5-iditan and $AH1$
[c]Isosorbide
[d]2,5-Mannitan

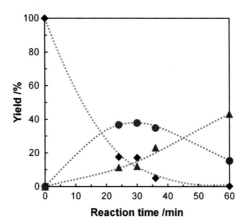

Fig. 5.3 Time course of sorbitol dehydration by Nafion SAC-13 at 403 K. Conditions: sorbitol 182 mg (1 mmol); Nafion SAC-13 50 mg. Legends: *black diamonds* amount of unreacted sorbitol; *red circles* yield of 1,4-sorbitan; and *blue triangles* yield of isosorbide. Each data point is the result of each batch reaction with different reaction times. *Dashed lines* are simply smooth lines for connecting experimental data

yields of 1,4-sorbitan (≤31 %, entries 7–9), and the colors of the catalysts also changed from white to brown or black during the reaction because of coke formation. SiO_2-Al_2O_3 was inactive in this reaction (entry 10). Among the catalysts tested, SZ afforded the highest yield and selectivity for the production of 1,4-sorbitan from sorbitol.

Since pristine ZrO_2 does not convert sorbitol (entry 4), acid species introduced by sulfation is responsible for the high catalytic activity of SZ. A reuse experiment of SZ was performed to evaluate the durability, resulting in a decrease of 1,4-sorbitan yield to 40 % in the second run (entry 3). The leaching amount of SO_4^{2-} corresponded to 0.4 % of SO_4^{2-} loaded on ZrO_2. In a control experiment, the equivalent amount of H_2SO_4 (1.0 μmol) was used for the dehydration of sorbitol (entry 11). Even such a small amount of H_2SO_4 has afforded 56 % yield of 1,4-sorbitan with 73 % selectivity, which are consistent with the results by pristine SZ (entry 2). Hence, the author concludes that leached species from SZ contribute to high selective synthesis of 1,4-sorbitan. As mentioned in Sect. 5.1, H_2SO_4 is currently used in an industrial process to produce 1,4-sorbitan due to its good catalytic activity [6, 7, 10]. However, the origins of good selectivity by H_2SO_4 remain unclear. Hence, in Sect. 5.3.3, the author elucidates characteristic activity of H_2SO_4 to obtain mechanistic insights, which would be a good basis of designing solid catalysts for selective synthesis of 1,4-sorbitan.

5.3.2 Identification of Reaction Products

Various dehydrated compounds are possibly formed from sorbitol, making identification of products difficult. The author isolated reaction products by HPLC equipped with a fraction collector, and then they were identified by combining 1D and 2D NMR as well as LC/MS. Both ^{1}H NMR spectrum and LC/MS spectrum of 1,4-sorbitan produced from sorbitol were consistent with those of the commercial standard (see Figs. 7.6 and 7.7). Other products, i.e., isosorbide, 2,5-mannitan, and 2,5-iditan, were similarly identified (see Figs. 7.8, 7.9 and 7.10). The LC/MS and NMR analyses also revealed that the HPLC peak of 2,5-iditan overlapped with that of the mono-anhydrohexitol *AH1*. *AH1* is not either 3,6-sorbitan (3,6-anhydrosorbitol) or 1,4-galactan (1,4-anhydrogalactitol) due to the inconsistence of ^{1}H and ^{13}C NMR spectra [23]. The HPLC analyses showed the formation of *AH2* (see Fig. 7.11). In comparison with retention times of commercial anhydrohexitols, neither *AH1* nor *AH2* was identified as 1,5-sorbitan or 1,5-mannitan. In addition to these compounds, minor products (<1 % yield) were also detected in HPLC, but they were not further characterized. The yields of 1,4-sorbitan, isosorbide, and 2,5-mannitan were calculated by HPLC with an absolute calibration method, in which their calibration factors were created using commercially available standard samples. For 2,5-iditan, *AH1*, and *AH2*, the standard samples are not commercially available, and thus the calibration factor of 1,4-sorbitan was applied to the calculation of their yields since regio- or stereoisomers generally give similar calibration factors in RI measurement.

5.3.3 Kinetic Study of Sorbitol Dehydration Catalyzed by H_2SO_4

To clarify the origins of good catalytic performance of H_2SO_4, kinetic study was performed for sorbitol dehydration in the presence of H_2SO_4 at 403 K (Fig. 5.4), in which larger amount of the acid (6.5 μmol) than the previous test (Table 5.1, entry 11) was employed to enhance reaction rates. The amount of unreacted sorbitol was on the decrease to 7.4 % in the linear function from 0 to 45 min, and then the consumption rate lessened at low amounts of remaining sorbitol. This manner is derived from the change of apparent reaction order from zero to first. In other words, this dehydration reaction involves pre-equilibrium, i.e., association equilibrium between the substrate and H_2SO_4. While sorbitol was consumed, the yield of 1,4-sorbitan steadily increased to 58 % until 45 min, followed by the gradual decrease of yield. Although 1,4-sorbitan successively underwent dehydration to form isosorbide, this reaction hardly occurred in the initial 30 min regardless of accumulation of 1,4-sorbitan in the reaction system. After exhausting almost all sorbitol, the dehydration of 1,4-sorbitan to isosorbide remarkably proceeded and the yield of isosorbide reached 72 % in 180 min. These time courses imply that the

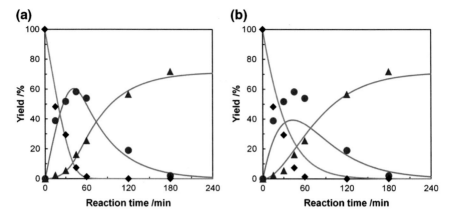

Fig. 5.4 Time course of sorbitol dehydration by H_2SO_4 at 403 K. Conditions: sorbitol 1 mmol; H_2SO_4 6.5 μmol. *Data points* show experimental results and *lines* are simulation curves based on Eqs. 5.4–5.9. **a** The curves were drawn by considering preferential association of H_2SO_4 with sorbitol over 1,4-sorbitan and isosorbide (i.e., $K_1/K_2 = 4.3$ and $K_1/K_3 = 4.5$); in contrast, **b** the preferential association was ignored (e.g., $K_1 = K_2 = K_3$). Legends: *black diamonds* amount of unreacted sorbitol; *red circles* yield of 1,4-sorbitan; and *blue triangles* yield of isosorbide. For **a**, the standard deviations for the difference between experimental data and simulation data were calculated to be 3.9 % for sorbitol, 2.2 % for 1,4-sorbitan, and 1.1 % for isosorbide

association of H_2SO_4 with sorbitol is more favorable than that with 1,4-sorbitan; that is, the dehydration of 1,4-sorbitan would be inhibited in the presence of sorbitol.

For quantitative evaluation of the reaction, the time course was simulated by assuming the Michaelis-Menten kinetic system including both pre-equilibrium of association and subsequent dehydration steps (see the scheme in Table 5.2). k_1 and k_2 are rate constants for dehydration of sorbitol and 1,4-sorbitan, respectively, and k_3, k_4, and k_5 are those for side reactions of sorbitol, 1,4-sorbitan, and isosorbide. The rates of side reactions were calculated from carbon balance of sorbitol, 1,4-sorbitan, and isosorbide. The association equilibrium constants for H_2SO_4 with sorbitol, 1,4-sorbitan, and isosorbide are represented as K_1, K_2, and K_3, respectively. The equilibrium constants are defined as Eqs. 5.1–5.3.

$$K_1 = \frac{[\text{Sorbitol@}H_2SO_4]}{[\text{Sorbitol}]\big([H_2SO_4]_0 - [\text{Sorbitol@}H_2SO_4] - [\text{Sorbitan@}H_2SO_4] - [\text{Isosorbide@}H_2SO_4]\big)}$$

$$(5.1)$$

$$K_2 = \frac{[\text{Sorbitan@}H_2SO_4]}{[\text{Sorbitan}]\big([H_2SO_4]_0 - [\text{Sorbitol@}H_2SO_4] - [\text{Sorbitan@}H_2SO_4] - [\text{Isosorbide@}H_2SO_4]\big)}$$

$$(5.2)$$

Table 5.2   Rate constants and ratios of association equilibrium constants for sorbitol dehydration by H_2SO_4 at 403 K

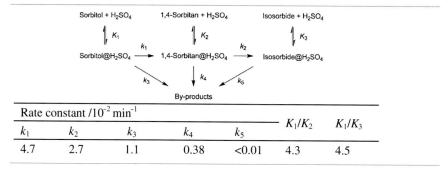

Rate constant $/10^{-2}$ min^{-1}					K_1/K_2	K_1/K_3
k_1	k_2	k_3	k_4	k_5		
4.7	2.7	1.1	0.38	<0.01	4.3	4.5

$$K_3 = \frac{[\text{Isosorbide@H}_2\text{SO}_4]}{[\text{Isosorbide}]\left([\text{H}_2\text{SO}_4]_0 - [\text{Sorbitol@H}_2\text{SO}_4] - [\text{Sorbitan@H}_2\text{SO}_4] - [\text{Isosorbide@H}_2\text{SO}_4]\right)} \tag{5.3}$$

where $[H_2SO_4]_0$ (unit: M) is initial concentration of H_2SO_4. [Sorbitol], [Sorbitan], and [Isosorbide] (M) are concentrations of sorbitol, 1,4-sorbitan, and isosorbide when reaching equilibrium. [Sorbitol@H$_2$SO$_4$], [Sorbitan@H$_2$SO$_4$], and [Isosorbide@H$_2$SO$_4$] (M) are concentrations of H_2SO_4 adducts with sorbitol, 1,4-sorbitan, and isosorbide when reaching equilibrium. Accordingly, the concentrations of H_2SO_4 adducts can be estimated from Eqs. 5.4 to 5.6

$$[\text{Sorbitol@H}_2\text{SO}_4] = \frac{K_1[\text{Sorbitol}][\text{H}_2\text{SO}_4]_0}{1 + K_1[\text{Sorbitol}] + K_2[\text{Sorbitan}] + K_3[\text{Isosorbide}]} \tag{5.4}$$

$$[\text{Sorbitan@H}_2\text{SO}_4] = \frac{K_2[\text{Sorbitan}][\text{H}_2\text{SO}_4]_0}{1 + K_1[\text{Sorbitol}] + K_2[\text{Sorbitan}] + K_3[\text{Isosorbide}]} \tag{5.5}$$

$$[\text{Isosorbide@H}_2\text{SO}_4] = \frac{K_3[\text{Isosorbide}][\text{H}_2\text{SO}_4]_0}{1 + K_1[\text{Sorbitol}] + K_2[\text{Sorbitan}] + K_3[\text{Isosorbide}]} \tag{5.6}$$

The reaction rates corresponding to each reactant are represented using reaction time t (unit: min^{-1}) as

$$\frac{d[\text{Sorbitol}]}{dt} = -(k_1 + k_3)[\text{Sorbitol@H}_2\text{SO}_4] \tag{5.7}$$

$$\frac{d[\text{Sorbitan}]}{dt} = k_1[\text{Sorbitol@H}_2\text{SO}_4] - (k_2 + k_4)[\text{Sorbitan@H}_2\text{SO}_4] \tag{5.8}$$

$$\frac{d[\text{Isosorbide}]}{dt} = k_2[\text{Sorbitan@H}_2\text{SO}_4] - k_5[\text{Isosorbide@H}_2\text{SO}_4] \tag{5.9}$$

in which the volume of reaction system at 403 K is 1.32×10^{-4} L (0.182 g $\times 0.726 \times 10^{-3}$ L g^{-1} [24]), which is used for the simulation of each concentration.

The lines in Fig. 5.4a shows the resulting simulation curves based on Eqs. 5.4–5.9 and fit well with the experimental data (dots in the figure). According to the simulation, the rate constants k_1 and k_2 were determined to be 4.7×10^{-2} and 2.7×10^{-2} min^{-1}, respectively (Table 5.2). The dehydration of sorbitol proceeded only 1.7 times faster than that of 1,4-sorbitan; this difference slightly contributes to the selective synthesis of 1,4-sorbitan from sorbitol in the presence of H$_2$SO$_4$. The side reactions of sorbitol and 1,4-sorbitan were minor due to their low rate constants ($k_3 = 1.1 \times 10^{-2}$ min^{-1} and $k_4 = 3.8 \times 10^{-3}$ min^{-1}), which were at least 2.5 times lower than k_1 and k_2. Indeed, the yields of by-products, e.g., 2,5-mannitan, 2,5-iditan, *AH1*, and *AH2*, were low. The rate constant k_5 was $<1.0 \times 10^{-4}$ min^{-1}, and thus isosorbide is highly thermally stable.

The simulation also estimated the ratios of K_1/K_2 and K_1/K_3 to be 4.3 and 4.5, respectively. It is worth noting that the absolute value for each equilibrium constant is unable to be elucidated since the simulation based on Eqs. 5.4–5.9 does not depend on their absolute values but on the ratios of K_1/K_2 and K_1/K_3. These high ratios argue that H$_2$SO$_4$ associates with sorbitol 4.3 and 4.5 more strongly than with 1,4-sorbitan and isosorbide; that is, the adduct formation of H$_2$SO$_4$ with 1,4-sorbitan and isosorbide is hampered in the presence of sorbitol. If this preferential association did not occur (e.g., $K_1 = K_2 = K_3$), the yield of 1,4-sorbitan would be at most 40 % by employing the same rate constants (Fig. 5.4b). In this case, the dehydration of sorbitol needs to proceed 3.2 times faster than conversion of 1,4-sorbitan in order to obtain the original yield (58 %); in other words, the ratio of $k_1/(k_2 + k_4)$ should be greater than 3.2. However, such a high value is difficult to be achieved since both dehydration reactions of sorbitol and 1,4-sorbitan are similarly catalyzed by H$_2$SO$_4$ regardless of pre-equilibrium as mentioned above. More serious issue in this case was that the simulation curves did not reproduce experimental data even when the rate constants were changed. For these results, the author has concluded that the preferential association of H$_2$SO$_4$ with sorbitol over 1,4-sorbitan should be included in the reaction and is a key for the selective production of 1,4-sorbitan. This conclusion is consistent with the contention by Takagaki et al. that molecular recognition plays an important role in the selective production of 1,4-sorbitan from sorbitol [25].

To design new catalysts based on these mechanistic insights, the author speculates that the combination of Brønsted acid sites with metal cation centers is useful for the selective formation of 1,4-sorbitan. Some metal cations such as Ca^{2+} and Pb^{2+} can form chelates as applied to HPLC columns for the separation of sugar compounds. For example, a commercially available HPLC column containing Pb^{2+} species retains sorbitol for a long time (41 min, see Fig. 7.11b, c) but does not

1,4-sorbitan (15 min). If Brønsted acid sites are juxtaposed with the chelating sites, 1,4-sorbitan may be selectively produced from sorbitol.

5.3.4 Reaction Mechanism for Dehydration of Sorbitol Catalyzed by H_2SO_4

Kinetic study discussed in Sect. 5.3.3 has revealed that sorbitol initially associates with H_2SO_4 to be activated prior to dehydration reaction. In this section, the dehydration mechanism is deduced from the reaction results and product identification mentioned in the previous sections. As shown in Fig. 5.5, five types of

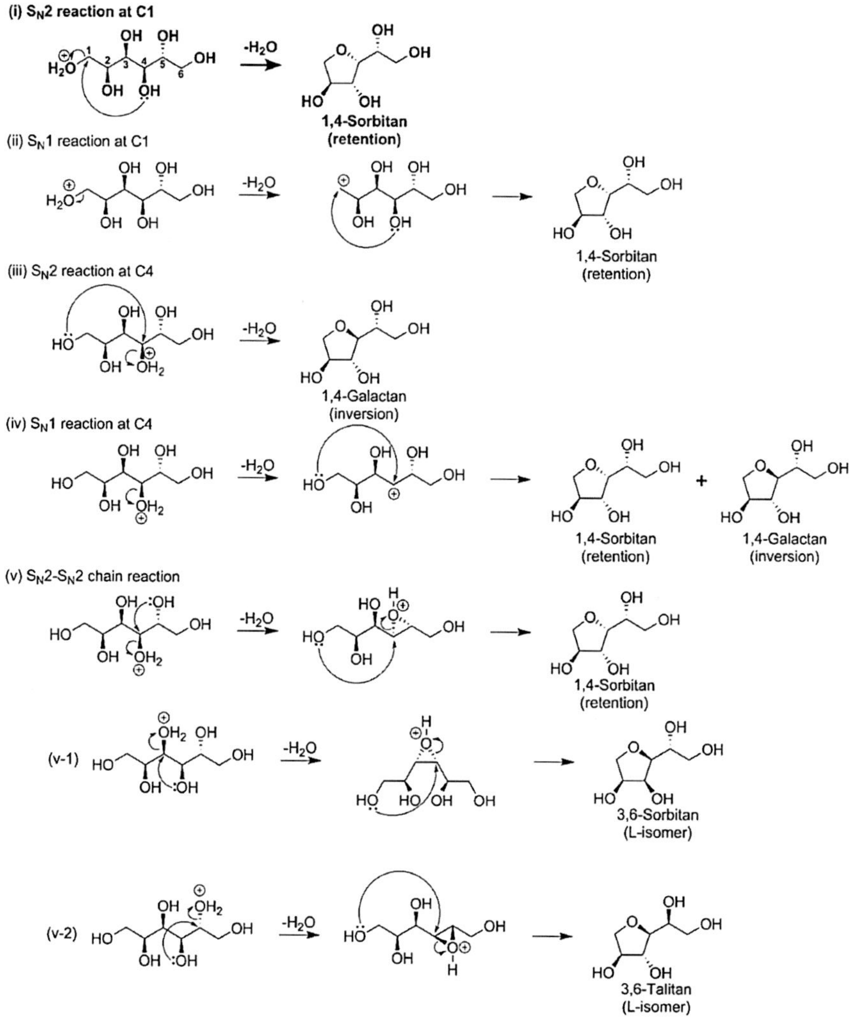

Fig. 5.5 Possible reaction mechanisms for 1,4-dehydration of sorbitol

reaction mechanism are presumable for the 1,4-dehydration of sorbitol: (i) S_N2 reaction at primary C1 attacked by an OH of C4 [26], (ii) S_N1 reaction at C1 attacked by the OH of C4, (iii) S_N2 reaction at secondary C4 attacked by an OH of C1, (iv) S_N1 reaction at C4 attacked by the OH of C1, and (v) S_N2-S_N2 chain reactions caused by a vicinal OH and terminal OH [27]. The author proposes that the 1,4-dehydration of sorbitol proceeds through the mechanism (i). The primary C1 easily undergoes the nucleophilic attack by the OH of C4 in the S_N2 manner owing to a low steric hindrance, resulting in the retention of stereochemistry of sorbitol, i.e., the formation of 1,4-sorbitan. This mechanism is consistent with the reaction result that 1,4-sorbitan has been selectively obtained. In contrast, the mechanism (ii) provides an energetically unfavorable primary carbocation as an intermediate, which should be immediately transformed into more stable secondary carbocation via 1,2-hydride shift. In a computation work regarding dehydration of sugar molecules, secondary carbocation is readily formed via 1,2-hydride shift even though primary carbocation is produced as an initial intermediate [28]. As a result, cyclic sugar alcohols other than 1,4-sorbitan should be major products, and thus the mechanism (ii) is not included in this study. In the mechanism (iii), the stereochemistry of C4 is changed from R to S due to the S_N2 reaction, resulting in the formation of 1,4-galactan without that of 1,4-sorbitan. 1,4-Galactan has not been observed in the reaction cocktails, and this mechanism (iii) is excluded. The mechanism (iv) is also negligible since this mechanism provides a secondary carbocation intermediate followed by the formation of 1,4-sorbitan as well as 1,4-galactan. In the mechanism (v), if the first S_N2 reaction occurs at C4 attacked by OH of C3 or C5, the stereochemistry of sorbitol is maintained to produce 1,4-sorbitan as shown in Fig. 5.3 However, the similar S_N2 reactions are also possible at C3 [mechanism (v-1)] and C5 [mechanism (v-2)] attacked by OH of C4, leading to the formation of 3,6-sorbitan (L-isomer) and 3,6-talitan (3,6-anhydrotalitol, L-isomer). However, these compounds were not detected. The author has concluded that the selective 1,4-dehydration of sorbitol by H_2SO_4 proceeds through the mechanism (i), i.e., the S_N2 reaction at C1 attacked by OH at C4.

5.4   Conclusions

In sorbitol dehydration, SZ affords 1,4-sorbitan in 60 % yield with 62 % selectivity; however, actual active species are not acids immobilized on SZ, but leached ones. The control reaction using 1.0 μmol of H_2SO_4, the amount of which is equivalent to that of leached species, gives similar yield and selectivity of 1,4-sorbitan. In other words, H_2SO_4 shows characteristic performance for selective synthesis of 1,4-sorbitan from sorbitol.

Kinetic studies of sorbitol dehydration by H_2SO_4 were conducted to understand the origins of its good catalytic performance. The results show that the high ratio of K_1/K_2 (4.3) mainly contributes to the good yield and selectivity of 1,4-sorbitan in addition to the small difference in rate constants ($k_1/k_2 = 1.7$). The association of

1,4-sorbitan with H_2SO_4 is substantially less favored in the presence of sorbitol, suppressing further dehydration of 1,4-sorbitan to isosorbide. Therefore, control of association equilibrium constant (K) is effective as well as that of reaction rate (k) to maximize the yield of 1,4-sorbitan. This insight is useful to the design of new catalysts for selective synthesis of anhydropolyols.

References

1. Werpy T, Petersen G, Aden A, Bozell J, Holladay J, White J, Manheim A, Eliot D, Lasure L, Jones S, Gerber M, Ibsen K, Lumberg L, Kelley S. Top value added chemicals from biomass. Volume I: results of screening for potential candidates from sugars and synthesis gas. (Online) http://www.dtic.mil/cgi-bin/GetTRDoc?Location=U2&doc=GetTRDoc.pdf&AD= ADA436528 Accessed 31 Oct 2015
2. Kobayashi H, Fukuoka A (2013) Synthesis and utilisation of sugar compounds derived from lignocellulosic biomass. Green Chem 15(7):1740–1763
3. Yabushita M, Kobayashi H, Fukuoka A (2014) Catalytic transformation of cellulose into platform chemicals. Appl Catal B Environ 145:1–9
4. Li X, Su Y, Zhou X, Mo X (2009) Distribution of sorbitan monooleate in poly (L-lactide-co-ε-caprolactone) nanofibers from emulsion electrospinning. Colloids Surf B Biointerfaces 69 (2):221–224
5. Yoshioka T, Sternberg B, Florence AT (1994) Preparation and properties of vesicles (niosomes) of sorbitan monoesters (Span 20, 40, 60 and 80) and a sorbitan triester (Span 85). Int J Pharm 105(1):1–6
6. Soltzberg S (1945) Sorbitan and process for making the same. US Patent 2,390,395
7. Soltzberg S, Goepp RM Jr, Freudenberg W (1946) Hexitol anhydrides. Synthesis and structure of arlitan, the 1,4-monoanhydride of sorbitol. J Am Chem Soc 68(6):919–921
8. Defaye J, Gadelle A, Pedersen C (1990) Acyloxonium ions in the high-yielding synthesis of oxolanes from alditols, hexoses, and hexonolactones catalysed by carboxylic acids in anhydrous hydrogen fluoride. Carbohydr Res 205:191–202
9. Montassier C, Ménézo J, Moukolo J, Naja J, Hoang LC, Barbier J, Boitiaux JP (1991) Polyol conversions into furanic derivatives on bimetallic catalysts: Cu–Ru, Cu–Pt and Ru–Cu. J Mol Cat 70(1):65–84
10. Andrews MA, Bhatia KK, Fagan PJ (2004) Process for the manufacture of anhydro sugar alcohols with the assistance of a gas purge. US Patent 6,689,892
11. Moore KM, Sanborn AJ, Bloom P (2008) Process for the production of anhydrosugar alcohols. US Patent 7,439,352
12. Sanborn AJ (2008) Process for the production of anhydrosugar alcohols. US Patent 7,420,067
13. Gu M, Yu D, Zhang H, Sun P, Huang H (2009) Metal (IV) phosphates as solid catalysts for selective dehydration of sorbitol to isosorbide. Catal Lett 133(1):214–220
14. Holladay JE, Hu J, Zhang X, Wang Y (2010) Methods for dehydration of sugars and sugar alcohols. US Patent 7,772,412
15. Xia J, Yu D, Hu Y, Zou B, Sun P, Li H, Huang H (2011) Sulfated copper oxide: an efficient catalyst for dehydration of sorbitol to isosorbide. Catal Commun 12(6):544–547
16. Liu A, Luckett CC (2011) Sorbitol conversion process. US Patent 7,982,059
17. Yamaguchi A, Hiyoshi N, Sato O, Shirai M (2011) Sorbitol dehydration in high temperature liquid water. Green Chem 13(4):873–881
18. Khan NA, Mishra DK, Ahmed I, Yoon JW, Hwang J-S, Jhung SH (2013) Liquid-phase dehydration of sorbitol to isosorbide using sulfated zirconia as a solid acid catalyst. Appl Catal A Gen 452:34–38

19. Ahmed I, Khan NA, Mishra DK, Lee JS, Hwang J-S, Jhung SH (2013) Liquid-phase dehydration of sorbitol to isosorbide using sulfated titania as a solid acid catalyst. Chem Eng Sci 93:91–95

20. Kobayashi H, Yokoyama H, Feng B, Fukuoka A (2015) Dehydration of sorbitol to isosorbide over H-beta zeolites with high Si/Al ratios. Green Chem 17(5):2732–2735

21. Reddy BM, Patil MK (2009) Organic syntheses and transformations catalyzed by sulfated zirconia. Chem Rev 109(6):2185–2208

22. Matsuhashi H, Nakamura H, Ishihara T, Iwamoto S, Kamiya Y, Kobayashi J, Kubota Y, Yamada T, Matsuda T, Matsushita K, Nakai K, Nishiguchi H, Ogura M, Okazaki N, Sato S, Shimizu K, Shishido T, Yamazoe S, Takeguchi T, Tomishige K, Yamashita H, Niwa M, Katada N (2009) Characterization of sulfated zirconia prepared using reference catalysts and application to several model reactions. Appl Catal A Gen 360(1):89–97

23. Wieneke R, Klein S, Geyer A, Loos E (2007) Structural and functional characterization of galactooligosaccharides in *Nostoc commune*: β-D-galactofuranosyl-(1→6)-[β-D-galactofuranosyl-(1→6)]₂-β-D-1,4-anhydrogalactitol and β-(1→6)-galactofuranosylated homologues. Carbohydr Res 342(18):2757–2765

24. Naoki M, Ujita K, Kashima S (1993) Pressure-volume-temperature relations and configurational energy of liquid, crystal, and glasses of D-sorbitol. J Phys Chem 97 (47):12356–12362

25. Morita Y, Furusato S, Takagaki A, Hayashi S, Kikuchi R, Oyama ST (2014) Intercalation-controlled cyclodehydration of sorbitol in water over layered-niobium-molybdate solid acid. ChemSusChem 7(3):748–752

26. Došen-Mićović L, Čeković Ž (1998) Conformational effects on the mechanism of acid-catalyzed dehydration of hexitols. J Phys Org Chem 11(12):887–894

27. Barker R (1970) Conversion of acyclic carbohydrates to tetrahydrofuran derivatives. Acid-catalyzed dehydration of hexitols. J Org Chem 35(2):461–464

28. Nimlos MR, Blanksby SJ, Qian X, Himmel ME, Johnson DK (2006) Mechanisms of glycerol dehydration. J Phys Chem A 110(18):6145–6156

Chapter 6
General Conclusions

In this thesis work, the author has demonstrated catalytic and mechanistic study on conversion of non-food and abundant biomass resources, cellulose and chitin. In Chap. 1, the reported systems for cellulose and chitin conversion including laboratory and industrial scales are summarized and discussed to clarify the current issues in biorefinery.

The important finding in Chap. 2 is that carbon materials bearing only weakly acidic functional groups catalyze hydrolysis of cellulose. For practical applications, the weakly acidic carbons are advantageous over reported sulfonated catalysts owing to their high hydrothermal stability and high resistance against ion exchange. Additionally, a new pretreatment method named *mix-milling* has been developed to overcome a serious issue of solid-solid reaction, specifically loose contact between solids. The mix-milling makes tight contact and drastically enhances the rate of hydrolysis of solid cellulose by solid carbons. As a result, 88 % yield of glucose with 90 % selectivity has been achieved under the optimized conditions, and this is one of the highest yields ever reported. The mix-milling pretreatment is also applicable to hydrolysis of raw biomass, bagasse kraft pulp.

Mechanistic studies to reveal the catalysis of carbon materials for cellulose hydrolysis have been conducted in Chap. 3. The first adsorption process of cellulosic molecules on carbon surface is driven by hydrophobic functionalities, namely formation of CH–π hydrogen bonds in addition to hydrophobic interactions. Following the adsorption, glycosidic bonds are activated and hydrolyzed by vicinal hydrophilic functional groups, the structures of which are similar to those of salicylic acid and phthalic acid. The vicinal groups accelerate the hydrolysis by increasing the occasion to interact with glycosidic bonds. Based on the mechanistic insights, the author has proposed the reaction mechanism and has also demonstrated the design of more active carbon catalysts, oxidized zeolite-templated carbon.

Two-step depolymerization of chitin, the most abundant nitrogen-containing biomass resource, to monomers with preserving N-acetyl groups has been demonstrated in Chap. 4. Mechanical force assisted the first depolymerization step of chitin to water-soluble oligomers in the presence of H_2SO_4. The produced oligomers show higher reactivity than original chitin and easily undergo further solvolysis to form N-acetylated monomers. The hydrolysis and methanolysis afford N-acetylglucosamine (GlcNAc) and 1-O-methyl-N-acetylglucosamine

© Springer Science+Business Media Singapore 2016
M. Yabushita, *A Study on Catalytic Conversion of Non-Food Biomass into Chemicals*, Springer Theses, DOI 10.1007/978-981-10-0332-5_6

(MeGlcNAc) in good yields of 53 and 70 %, respectively. During these reactions, N-acetyl groups in the substrates are almost completely retained. This two-step catalytic reaction system enables to reduce the use of acid by 99.8 % compared to the conventional processes.

In Chap. 5, the selective dehydration of sorbitol to 1,4-sorbitan catalyzed by H_2SO_4 has been investigated to reveal its catalysis. In general, 1,4-sorbitan readily undergoes successive dehydration to isosorbide by acid catalysts. However, kinetic study has indicated that the transformation of 1,4-sorbitan by H_2SO_4 is hampered in the presence of sorbitol since H_2SO_4 preferentially associates with sorbitol over 1,4-sorbitan. As a result, H_2SO_4 provides good yield and selectivity of 1,4-sorbitan. A blueprint for designing new catalysts based on the mechanistic studies has been proposed.

In conclusion, the author has achieved the development of new catalytic systems to produce useful platform chemicals, i.e., glucose, GlcNAc, MeGlcNAc, and 1,4-sorbitan, from cellulose and chitin in good yields. These catalytic systems are promising for building up sustainable societies since produced value-added chemicals are potential alternatives to petroleum in a variety of fields, specifically highly functionalized chemicals such as medicines and polymers. Besides, the results in this work clearly show that both catalyst design and reaction engineering are extremely effective for high-yielding production of chemicals from biomass resources, and this manner is useful in the next-generation biorefinery.

Nowadays, in the field of catalytic biomass transformation, a number of researchers have achieved high-yielding production of valuable chemicals by simple improvements and/or combinations of known catalytic systems. In other words, only a few researches show new concepts of reactions and catalysts, as well as *basic chemistry*. This fashion limits further applications of reported catalysts because their characteristics and advantages remain as missing pieces of the puzzle. Herein, this thesis work shows both basic catalytic chemistry and its applications in detail. The author strongly believes that the combination of basic and application chemistry will cause paradigm shifts in biorefinery.

Chapter 7
Appendices

In this chapter, the author has summarized the following discussion and data: (i) theoretical estimation of adsorption entropy change for cellobiose adsorption on carbon surface (Chap. 3); (ii) identification of chitin oligomers (Chap. 4); (iii) identification of MeGlcNAc (Chap. 4); and (iv) identification of reaction produces of sorbitol dehydration (Chap. 5).

7.1 Theoretical Estimation of Adsorption Entropy Change for Cellobiose Adsorption on Carbon Surface

The entropy change of cellobiose adsorption on carbon surface (ΔS°_{ads}) is theoretically calculated in this section. ΔS°_{ads} is divided into nine constituents (Eq. 7.1).

$$\Delta S^\circ_{ads} = \Delta S^\circ_{hydrophobic}$$
$$+ \Delta S^\circ_{trans}(Cellobiose) + \Delta S^\circ_{rot}(Cellobiose) + \Delta S^\circ_{vib}(Cellobiose) + \Delta S^\circ_{conf}(Cellobiose)$$
$$+ \Delta S^\circ_{trans}(Carbon) + \Delta S^\circ_{rot}(Carbon) + \Delta S^\circ_{vib}(Carbon) + \Delta S^\circ_{conf}(Carbon)$$

$$(7.1)$$

where $\Delta S^\circ_{hydrophobic}$ is entropy change of hydrophobic interaction. $\Delta S^\circ_{trans}(Cellobiose)$, $\Delta S^\circ_{rot}(Cellobiose)$, $\Delta S^\circ_{vib}(Cellobiose)$, and $\Delta S^\circ_{conf}(Cellobiose)$ are entropy changes of translation, rotation, vibration, and conformation for cellobiose, respectively. $\Delta S^\circ_{trans}(Carbon)$, $\Delta S^\circ_{rot}(Carbon)$, $\Delta S^\circ_{vib}(Carbon)$, and $\Delta S^\circ_{conf}(Carbon)$ are entropy changes of translation, rotation, vibration, and conformation for carbon material. For calculation, the author has assumed that $\Delta S^\circ_{trans}(Carbon)$ and $\Delta S^\circ_{rot}(Carbon)$ are negligible since the size of carbon material is significantly larger than cellobiose (see Figs. 3.16 and 3.19) and its motion would not be restricted by cellobiose adsorption. $\Delta S^\circ_{vib}(Cellobiose)$ and $\Delta S^\circ_{vib}(Carbon)$ are approximately zero, as almost all kinds of molecular vibrations are in ground state *zero-point vibration* (viz., $S^\circ_{vib} = 0$) at room temperature [1]. Furthermore, $\Delta S^\circ_{conf}(Cellobiose)$ and $\Delta S^\circ_{conf}(Carbon)$ would also be zero, as the DFT calculations have shown that conformations of cellobiose and carbon material consisting of quadruple-layered

© Springer Science+Business Media Singapore 2016
M. Yabushita, *A Study on Catalytic Conversion of Non-Food Biomass into Chemicals*, Springer Theses, DOI 10.1007/978-981-10-0332-5_7

graphene sheets are retained during adsorption (see Fig. 3.22). Accordingly, $\Delta S°_{hydrophobic}$, $\Delta S°_{trans}$(Cellobiose), and $\Delta S°_{rot}$(Cellobiose) values are necessary for the determination of theoretical $\Delta S°_{ads}$ value. First, for $\Delta S°_{hydrophobic}$, it is reported that a standard entropy of +4.2 J K^{-1} mol^{-1} increases when one water molecule is desorbed from a substrate molecule [2]. The adsorption cross-sectional area of one cellobiose molecule is 0.8 nm^2 as shown in Fig. 3.19, since cellobiose adsorbs on carbon surface by its axial plane to form CH–π hydrogen bonds. This explains that 15–24 water molecules exist on one axial plane of cellobiose. The same number of water molecules on a carbon surface is also removed in the adsorption due to hydrophobic interactions; in total, 30–48 water molecules gain an entropy increase. Therefore, the total entropy increase by hydrophobic interaction ($\Delta S°_{hydrophobic}$) is +164 ± 37 J K^{-1} mol^{-1}. Next, $\Delta S°_{trans}$(Cellobiose) and $\Delta S°_{rot}$(Cellobiose) can be elucidated by the Sackur–Tetrode equation (Eq. 7.2) and Eq. 7.3, respectively [1]. The number of 1/3 in Eq. 7.2 is derived from a restriction of molecular motion. Free molecules can move in 3D space, whereas adsorbed molecules can move on 2D surface; hence, adsorbed molecules lose one of the three translational DFs. The calculated value in the brackets followed by 1/3 in Eq. 7.2 gives whole translational entropy, and thus the multiplication of the value with 1/3 is required to estimate translational entropy change in adsorption. For the same reason, two rotational DFs are lost on 2D surface (in 3D space, molecules have three rotational DFs), and the value of 2/3 is included in Eq. 7.3. The author should note that these equations are for ideal gas and the values obtained from these are overestimated for real systems [1].

$$\Delta S°_{trans}(\text{Cellobiose}) = \frac{1}{3}\left\{ R\ln\left[\left(\frac{2\pi R}{h^2}\right)^{\frac{3}{2}} \cdot \frac{V}{N_A} \cdot e^{\frac{5}{2}} \right] + \frac{3}{2}R\ln M + \frac{3}{2}R\ln T \right\} \quad (7.2)$$

$$\Delta S°_{rot}(\text{Carbon}) = \frac{2}{3}\left\{ R\ln\left[\pi^{\frac{1}{2}}\left(\frac{8\pi^2 Re}{N_A h^2}\right)^{\frac{3}{2}} \right] + \frac{1}{2}R\ln(I_A I_B I_C) + \frac{3}{2}R\ln T \right\} \quad (7.3)$$

where R is the gas constant, h is the Planck constant, V is the volume of solution (in this calculation, V is 1×10^{-3} m^3), N_A is the Avogadro constant, e is the Napier's constant, M (unit: kg mol^{-1}) is the molecular weight, T (K) is temperature, and I_A, I_B, and I_C are principal moments of inertia. I_A, I_B, and I_C are defined by Eqs. 7.4, 7.5, and 7.6.

$$I_A = \frac{M}{12N_A}(a^2 + b^2) \quad (7.4)$$

$$I_B = \frac{M}{12N_A}(b^2 + c^2) \quad (7.5)$$

$$I_C = \frac{M}{12 N_A} (c^2 + a^2) \tag{7.6}$$

in which a, b, and c represent size of adsorbate; in this study, a, b, and c are 1.1, 0.7, and 0.4 nm, respectively (see Fig. 3.19).

Based on Eqs. 7.2–7.6, ΔS°_{trans}(Cellobiose) and ΔS°_{rot}(Cellobiose) at 296 K are calculated to be -51.6 and -99.5 J K^{-1} mol^{-1}. Consequently, the whole entropy change in adsorption (ΔS°_{ads}) is $+12.1 \pm 37$ J K^{-1} mol^{-1}. The number of water molecules related to hydrophobic interaction is not determined (30–48 molecules, vide supra), giving a wide range of this theoretical value. This theoretical value roughly agrees with the experimental one ($+23.5 \pm 3.4$ J K^{-1} mol^{-1}, Table 3.9, entry 52).

7.2 Identification of Chitin Oligomers

In this section, the author has summarized the raw data and the assignments of LC/MS (Fig. 7.1) and NMR spectroscopy (Fig. 7.2) of Chitin-H$_2$SO$_4$-PM. The NMR analyses indicate that the chitin oligomers contained in Chitin-H$_2$SO$_4$-PM have branched units linked by 1,6-glycosidic bond.

GlcNAc (N-acetylglucosamine, M = C$_8$H$_{15}$NO$_6$). LC/MS (*m/z*): 204 ([M + H – H$_2$O]$^+$), 222 ([M + H]$^+$).

(GlcNAc)$_2$ (dimer of GlcNAc, M = C$_{16}$H$_{28}$N$_2$O$_{11}$). LC/MS (*m/z*): 204 ([M + H – C$_8$H$_{15}$NO$_6$]$^+$), 407 ([M + H – H$_2$O]$^+$), 425 ([M + H)$^+$).

(GlcNAc)$_3$ (trimer of GlcNAc, M = C$_{24}$H$_{41}$N$_3$O$_{16}$). LC/MS (*m/z*): 204 ([M + H – C$_{16}$H$_{28}$N$_2$O$_{11}$]$^+$), 407 ([M + H – C$_8$H$_{15}$NO$_6$]$^+$), 610 ([M + H – H$_2$O]$^+$), 628 ([M + H]$^+$).

(GlcNAc)$_4$ (tetramer of GlcNAc, M = C$_{32}$H$_{54}$N$_4$O$_{21}$). LC/MS (*m/z*): 204 ([M + H – C$_{24}$H$_{41}$N$_3$O$_{16}$]$^+$), 407 ([M + H – C$_{16}$H$_{28}$N$_2$O$_{11}$]$^+$), 610 ([M + H – C$_8$H$_{15}$NO$_6$]$^+$), 813 ([M + H – H$_2$O]$^+$), 831 ([M + H]$^+$).

(GlcNAc)$_5$ (pentamer of GlcNAc, M = C$_{40}$H$_{67}$N$_5$O$_{26}$). LC/MS (*m/z*): 204 ([M + H – C$_{32}$H$_{54}$N$_4$O$_{21}$]$^+$), 407 ([M + H – C$_{24}$H$_{41}$N$_3$O$_{16}$]$^+$), 610 ([M + H – C$_{16}$H$_{28}$N$_2$O$_{11}$]$^+$), 813 ([M + H – C$_8$H$_{15}$NO$_6$]$^+$), 1016 ([M + H – H$_2$O]$^+$), 1034 ([M + H]$^+$).

(GlcNAc)$_6$ (hexamer of GlcNAc, M = C$_{48}$H$_{80}$N$_6$O$_{31}$). LC/MS (*m/z*): 204 ([M + H – C$_{40}$H$_{67}$N$_5$O$_{26}$]$^+$), 407 ([M + H – C$_{32}$H$_{54}$N$_4$O$_{21}$]$^+$), 610 ([M + H – C$_{24}$H$_{41}$N$_3$O$_{16}$]$^+$), 813 ([M + H – C$_{16}$H$_{28}$N$_2$O$_{11}$]$^+$), 1016 ([M + H – C$_8$H$_{15}$NO$_6$]$^+$), 1219 ([M + H – H$_2$O]$^+$), 1238 ([M + H]$^+$).

(GlcNAc)$_7$ (heptamer of GlcNAc, M = C$_{56}$H$_{93}$N$_7$O$_{36}$). LC/MS (*m/z*): 204 ([M + H – C$_{48}$H$_{80}$N$_6$O$_{31}$]$^+$), 407 ([M + H – C$_{40}$H$_{67}$N$_5$O$_{26}$]$^+$), 610 ([M + H – C$_{32}$H$_{54}$N$_4$O$_{21}$]$^+$), 813 ([M + H – C$_{24}$H$_{41}$N$_3$O$_{16}$]$^+$), 1016 ([M + H – C$_{16}$H$_{28}$N$_2$O$_{11}$]$^+$), 1220 ([M + H – C$_8$H$_{15}$NO$_6$]$^+$), 1422 ([M + H – H$_2$O]$^+$), 1442 ([M + H]$^+$).

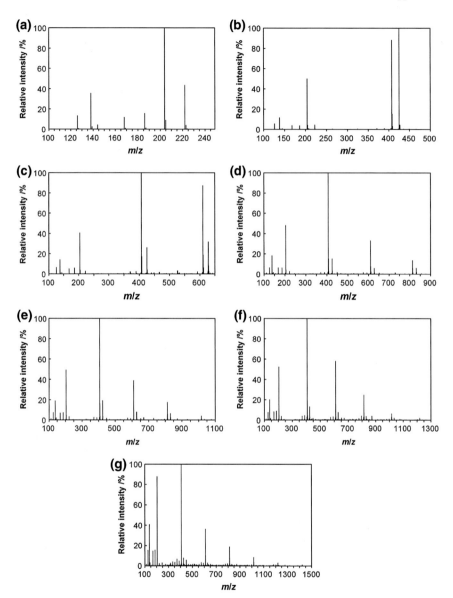

Fig. 7.1 LC/MS spectra of Chitin-H₂SO₄-PM, recorded by positive ion mode: **a** GlcNAc; **b** (GlcNAc)₂; **c** (GlcNAc)₃; **d** (GlcNAc)₄; **e** (GlcNAc)₅; **f** (GlcNAc)₆; and **g** (GlcNAc)₇. A SUGAR SH1011 column was used for the analyses

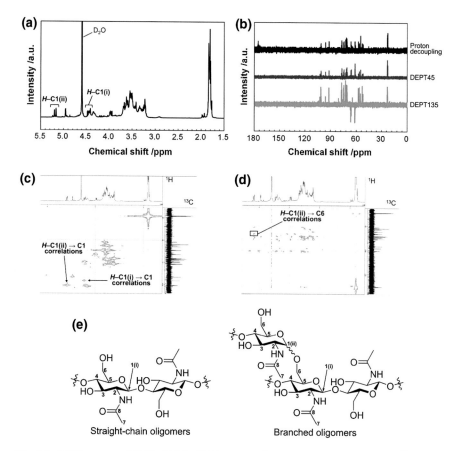

Fig. 7.2 NMR spectra of Chitin-H_2SO_4-PM in D_2O: **a** ^{1}H NMR; **b** proton-decoupled ^{13}C NMR and DEPT; **c** ^{13}C–^{1}H HMQC (horizontal axis: ^{1}H, vertical axis: ^{13}C); **d** ^{13}C–^{1}H HMBC (horizontal axis: ^{1}H, vertical axis: ^{13}C); and **e** possible structures of chitin oligomers contained in Chitin-H_2SO_4-PM

7.3 Identification of MeGlcNAc

Here the author has summarized the raw data and assignments of NMR spectroscopy (Fig. 7.3), IR spectroscopy (Fig. 7.4), and LC/MS (Fig. 7.5) of MeGlcNAc. For MeGlcNAc produced from chitin, elemental analysis was also performed (see below).

α-MeGlcNAc (1-O-methyl-α-N-acetylglucosamine, standard). ^{1}H NMR (400 MHz, D_2O): δ 4.74 (d, *J* = 3.6 Hz, 1H, *H*–C1), 3.90 (dd, *J* = 11.2, 3.6 Hz, 1H, *H*–C2), 3.87 (dd, *J* = 12.4, 2.4 Hz, 1H, *H*–C6), 3.77 (dd, *J* = 12.4, 5.6 Hz, 1H, *H*–C6), 3.70 (dd, *J* = 10.4, 9.2 Hz, 1H, *H*–C3), 3.66 (ddd, *J* = 9.6, 5.6, 2.4 Hz, 1H, *H*–C5), 3.46 (dd, *J* = 10.0, 9.2 Hz, 1H, *H*–C4), 3.37 (s, 3H, *H*–C9), 2.02 (s, 3H,

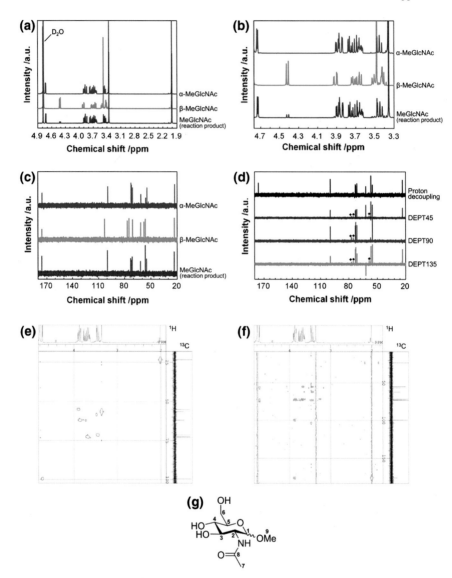

Fig. 7.3 NMR spectra of MeGlcNAc produced from Chitin-H$_2$SO$_4$-PM, recorded in D$_2$O: **a** ^{1}H NMR; **b** expanded view of the ^{1}H NMR spectum (3.30–4.77 ppm); **c** proton-decoupled ^{13}C NMR (comparison with standard samples); **d** proton-decoupled ^{13}C NMR and DEPT; **e** ^{13}C ^{1}H HMQC (horizontal axis: ^{1}H, vertical axis: ^{13}C); **f** ^{13}C ^{1}H HMBC (horizontal axis: ^{1}H, vertical axis: ^{13}C); and **g** structure of MeGlcNAc. Small peaks labeled as black diamonds in figure **d** are derived from β-MeGlcNAc. The ^{1}H NMR and DEPT spectra indicated that both α- and β-MeGlcNAc were produced by methanolysis of Chitin-H$_2$SO$_4$-PM. Based on the ^{1}H NMR spectrum, the ratio of α- to β-MeGlcNAc was determined to be 5.0

Fig. 7.4 IR spectra of
MeGlcNAc: **a** α-MeGlcNAc
(standard); **b** β-MeGlcNAc
(standard); and **c** MeGlcNAc
produced by methanolysis of
Chitin-H$_2$SO$_4$-PM.
Transmission mode, KBr disk

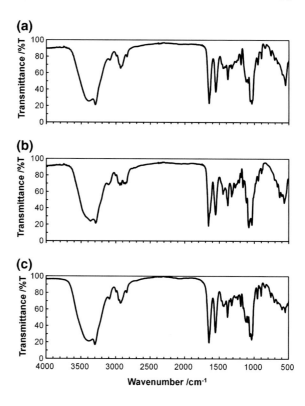

H–C7). ^{13}C NMR (100 MHz, D$_2$O): δ 175.4 (C, C8), 99.0 (CH, C1), 72.6 (CH,
C5), 72.1 (CH, C3), 70.9 (CH, C4), 61.5 (CH$_2$, C6), 56.1 (CH$_3$, C9), 54.5 (CH, C2),
22.8 (CH$_3$, C7). IR (KBr pellet, cm^{-1}): 3393 [ν(O–H)], 3296 [ν(N–H, amide)],
2803–3027 [ν(C–H, alkyl)], 2954 [ν(C–H, alkyl)], 2932 [ν(C–H, alkyl)], 2904
[ν(C–H, alkyl)], 1650 [ν(C=O, amide I)], 1554 [δ(N–H, amide II)]. LC/MS (*m/z*):
[M + H]$^+$ calculated 236; found 236.

β-MeGlcNAc   (1-O-methyl-β-N-acetylglucosamine,   standard). ^{1}H   NMR
(400 MHz, D$_2$O): δ 4.43 (d, *J* = 8.4 Hz, 1H, *H*–C1), 3.92 (dd, *J* = 12.4, 2.0 Hz, 1H,
H–C6), 3.73 (dd, *J* = 12.4, 5.4 Hz, *H*–C6), 3.67 (dd, *J* = 10.4, 8.8 Hz, *H*–C2), 3.53
(d, *J* = 8.0 Hz, *H*–C3), 3.49 (s, 3H, *H*–C9), 3.39–3.47 (m, 2H, *H*–C4 and *H*–C5),
2.02 (s, 3H, *H*–C7). ^{13}C NMR (100 MHz, D$_2$O): δ 175.7 (C, C8), 102.9 (CH, C1),
76.9 (CH, C5), 74.9 (CH, C3), 70.9 (CH, C4), 61.7 (CH$_2$, C6), 58.0 (CH$_3$, C9),
56.4 (CH, C2), 23.1 (CH$_3$, C7). IR (KBr pellet, cm^{-1}): 3370 [ν(O–H)], 3292 [ν(N–
H, amide)], 2790–3027 [ν(C–H, alkyl)], 1657 [ν(C=O, amide I)], 1554 [δ(N–H,
amide II)]. LC/MS (*m/z*): [M + H]$^+$ calculated 236; found 236.

*MeGlcNAc (1-O-methyl-N-acetylglucosamine, produced by methanolysis of
Chitin-H$_2$SO$_4$-PM).* Elemental analysis (calculated, found): C (45.95, 45.74), H
(7.28, 7.21), N (5.96, 5.96), O (40.81, 41.10). LC/MS (*m/z*): [M + H]$^+$ calculated
236; found 236. The NMR and IR spectra indicated that this MeGlcNAc is a
mixture of α- and β-MeGlcNAc in 5.0: 1 (Figs. 7.3, 7.4 and 7.5).

Fig. 7.5 LC/MS spectra of
MeGlcNAc, recorded by
positive ion mode:
a α-MeGlcNAc (standard);
b β-MeGlcNAc (standard);
and **c** MeGlcNAc produced
by methanolysis of
Chitin-H₂SO₄-PM.
A SUGAR SH1011 column
was used for the analyses

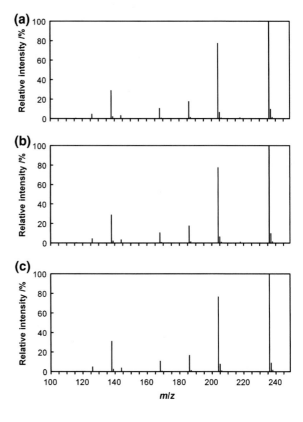

7.4   Identification of Reaction Products of Sorbitol Dehydration

This section includes the raw data and assignments of NMR spectroscopy, LC/MS, and HPLC of anhydrohexitols produced from sorbitol (Figs. 7.6, 7.7, 7.8, 7.9, 7.10, and 7.11).

Fig. 7.6 NMR spectra of
1,4-sorbitan in D₂O. All peaks
of 1,4-sorbitan produced from
sorbitol are consistent with
those of standard sample

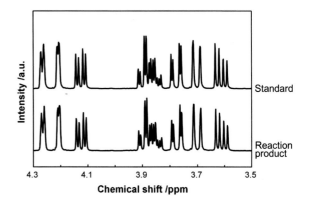

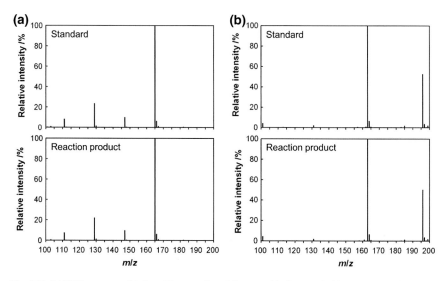

Fig. 7.7 LC/MS spectra of 1,4-sorbitan produced from sorbitol, recorded by **a** positive ion mode and **b** negative ion mode. A Rezex RPM-monosaccharide Pb++ column was used for the analyses

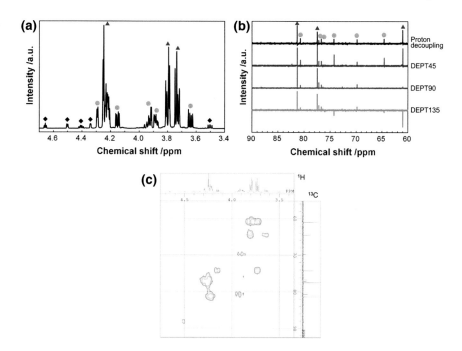

Fig. 7.8 NMR spectra of a reaction mixture containing isosorbide (*black diamonds*), 2,5-iditan (*blue triangles*), and *AH1* (*orange circles*): **a** ^{1}H NMR; **b** proton-decoupled ^{13}C NMR and DEPT; and **c** ^{13}C–^{1}H HMQC (horizontal axis: ^{1}H, vertical axis: ^{13}C). ^{1}H NMR peaks of 2,5-iditan are in agreement with those in a reference [3]. ^{1}H NMR peaks of isosorbide are consistent with those of a standard (not shown). ^{13}C NMR peaks of isosorbide enable to be distinguished from noise due to its low concentration

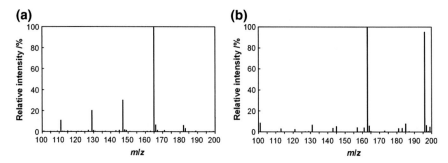

Fig. 7.9 LC/MS spectra of a mixture of isosorbide, 2,5-iditan, and *AH1*, recorded by **a** positive ion mode and **b** negative ion mode. A Rezex RPM-monosaccharide Pb++ column was used for the analyses. In figure **b**, [M − H]⁻ of isosorbide calculated to be 145 was not detected due to its low concentration and sensitivity

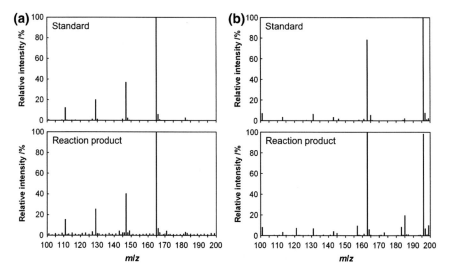

Fig. 7.10 LC/MS spectra of 2,5-mannitan, recorded by **a** positive ion mode and **b** negative ion mode. A Rezex RPM-Monosaccharide Pb++ column was used for the analyses. Several small peaks were observed from other products in trace quantities

1,4-Sorbitan (1,4-anhydrosorbitol). ^{1}H NMR (400 MHz, D$_2$O): δ 4.27 (1H, d, J = 4.0 Hz, *H*–C2), 4.21 (1H, d, J = 2.8 Hz, *H*–C3), 4.13 (1H, dd, J = 10, 4.0 Hz, *H*–C1), 3.90 (1H, dd, J = 8.8, 2.8 Hz, *H*–C4), 3.85 (1H, ddd, J = 8.8, 6.0, 2.8 Hz, *H*–C5), 3.78 (1H, dd, J = 12, 2.8 Hz, *H*–C6), 3.70 (1H, d, J = 10 Hz, *H*–C1), 3.61 (1H, dd, J = 12, 6.0 Hz, *H*–C6). *H*–C1 and *H*–C6 were distinguished by means of ^{1}H NMR in a DMSO-d$_6$ solvent. In DMSO-d$_6$, the peaks of *H*–C6 are observed as *ddd* due to coupling with one proton of the hydroxyl group in addition to two aliphatic protons, whereas the peaks of *H*–C1 are in *dd* form due to coupling with one proton of the hydroxyl group in addition to one aliphatic proton (not shown).

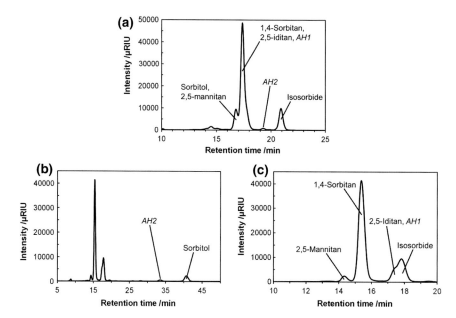

Fig. 7.11 HPLC charts for sorbitol dehydration by H_2SO_4, recorded by an RI detector: **a** a SUGAR SH1011 column; **b** a Rezex RPM-monosaccharide Pb++ column; and **c** an expanded chromatogram of figure **b**. Retention times for figure **a**: 2,5-mannitan (16.5 min); 1,5-sorbitan (16.5 min); sorbitol (16.6 min); 1,4-sorbitan, 2,5-iditan, and *AH1* (17–18 min); 1,5-mannitan (18.3 min); *AH2* (19.1 min); and isosorbide (20.9 min). Retention times for figure **b**: 2,5-mannitan (14.3 min); 1,5-sorbitan (15.3 min); 1,4-sorbitan (15.4 min); 2,5-iditan and *AH1* (17.5 min); isosorbide (17.8 min); 1,5-mannitan (21.8 min); *AH2* (33.5 min); and sorbitol (40.7 min)

LC/MS: [M + H]$^+$ calculated 165, found 165 (reaction product), found 165 (s-tandard); [M − H]$^-$ calculated 163, found 163 (reaction product), found 163 (standard).

Isosorbide (1,4:3,6-dianhydrosorbitol). ^{1}H NMR (400 MHz, D_2O): δ 4.64 (1H, dd, J = 5.2, 4.4 Hz, *H*–C4), 4.49 (1H, d, J = 4.4 Hz, *H*–C3), 4.39 (1H, ddd, J = 7.0, 7.0, 5.2 Hz, *H*–C5), 4.32 (1H, d, J = 3.2 Hz, *H*–C2), 3.89 3.96 (2H, *H*–C1 and *H*–C6), 3.87 (1H, dd, J = 10.4, 3.2 Hz, *H*–C1), 3.49 (1H, dd, J = 9.0, 7.2 Hz, *H*–C6). Some of the ^{1}H NMR peaks at 3.89 3.96 and 3.87 ppm were overlapped with the peaks of *AH1* (Fig. 7.8a). The ^{13}C NMR and DEPT peaks of isosorbide in Fig. 7.8b could not be distinguished from noise due to its low concentration.

2,5-Mannitan (2,5-anhydromannitol). LC/MS: [M + H]$^+$ calculated 165, found 165 (reaction product), found 165 (standard); [M − H]$^-$ calculated 163, found 163 (reaction product), found 163 (standard).

2,5-Iditan (2,5-anhydroiditol). ^{1}H NMR (600 MHz, D_2O): δ 4.21–4.25 (4H, *H*–C2, *H*–C3, *H*–C4, and *H*–C5), 3.80 (2H, dd, J = 11.4, 4.8 Hz, *H*–C1 and *H*–C6), 3.73 (2H, dd, J = 11.4, 7.2 Hz, *H*–C1 and *H*–C6). ^{13}C NMR (150 MHz, D_2O): δ 81.3 (CH, C3 and C4), 77.5 (CH, C2 and C5), 61.0 (CH_2, C1 and C6). The ^{1}H NMR peaks at 4.21–4.25, 3.80, and 3.73 ppm were overlapped with those of *AH1* (Fig. 7.8a). The assignment of the ^{13}C NMR peaks is based on a reference [3]. LC/MS: $[M + H]^+$ calculated 165, found 165; $[M - H]^-$ calculated 163, found 163. *AH1 (unidentified mono-anhydrohexitol 1).* ^{1}H NMR (600 MHz, D_2O): δ 4.29 (1H, d, *J* = 4.2 Hz), 4.15 (1H, dd, *J* = 10.5, 4.2 Hz), 3.92 (1H, dd, *J* = 9.0, 2.4 Hz), 3.88 (1H, ddd, *J* = 9.0, 6.0, 3.0 Hz), 3.64 (1H, dd, *J* = 12, 6.0 Hz). ^{13}C NMR (150 MHz, D_2O): δ 80.7 (CH), 77.2 (CH), 76.7 (CH), 74.3 (CH_2), 69.9 (CH), 64.6 (CH_2). Several ^{1}H NMR peaks were overlapped with those of 2,5-iditan and isosorbide (Fig. 7.8a). The chemical shifts of the ^{1}H NMR peaks indicate that this unidentified compound is not 3,6-sorbitan or 1,4-galactan [4]. Additionally, the HPLC analyses show that *AH1* is not 1,5-sorbitan or 1,5-mannitan (Fig. 7.11). LC/MS: $[M + H]^+$ found 165; $[M - H]^-$ found 163.

AH2 (unidentified mono-anhydrohexitol 2). The retention times of *AH2* in HPLC analyses disagree with those of 1,5-sorbitan and 1,5-mannitan (Fig. 7.11). LC/MS: $[M + H]^+$ found 165; $[M - H]^-$ found 163.

References

1. Mammen M, Shakhnovich EI, Deutch JM, Whitesides GM (1998) Estimating the entropic cost of self-assembly of multiparticle hydrogen-bonded aggregates based on the cyanuric acid·melamine lattice. J Org Chem 63(12):3821–3830
2. Southall NT, Dill KA, Haymet ADJ (2002) A view of the hydrophobic effect. J Phys Chem B 106(3):521–533
3. Bock K, Pedersen C, Thøgersen H (1981) Acid catalyzed dehydration of alditols. Part I. D-Glucitol and D-mannitol. Acta Chem Scand B 35(6):441–449
4. Wieneke R, Klein S, Geyer A, Loos E (2007) Structural and functional characterization of galactooligosaccharides in *Nostoc commune*: β-D-galactofuranosyl-(1→6)-[β-D-galactofuranosyl-(1→6)]₂-β-D-1,4-anhydrogalactitol and β-(1→6)-galactofuranosylated homologues. Carbohydr Res 342(18):2757–2765

Curriculum Vitae

1. Personal Information

Name: Mizuho Yabushita
Nationality: Japanese
Gender: Male
Current affiliation: (1) Institute for Catalysis, Hokkaido University
(2) Department of Chemical and Biomolecular Engineering,
University of California, Berkeley
E-mail: (1) m.yabushita@cat.hokudai.ac.jp
(2) m.yabushita@berkeley.edu

2. Education

Mar. 2011   Bachelor of Science
Faculty of Science, Hokkaido University, Sapporo, Japan
Sep. 2012   Master of Chemical Sciences and Engineering
Graduate School of Chemical Sciences and Engineering,
Hokkaido University, Sapporo, Japan
Mar. 2015   Doctor of Philosophy in Science
Graduate School of Chemical Sciences and Engineering,
Hokkaido University, Sapporo, Japan

© Springer Science+Business Media Singapore 2016
M. Yabushita, *A Study on Catalytic Conversion of Non-Food
Biomass into Chemicals*, Springer Theses, DOI 10.1007/978-981-10-0332-5

3. Professional Carrier

Apr. 2014–Mar. 2015   Research Fellow DC2
                       Japan Society for the Promotion of Science
Apr. 2015–Sep. 2015   Postdoctoral Fellow
                       Catalysis Research Center, Hokkaido University
Apr. 2015–   Research Fellow PD
                       Japan Society for the Promotion of Science
May 2015–   Visiting Scholar
                       Department of Chemical and Biomolecular Engineering,
                       University of California, Berkeley
Oct. 2015–   Postdoctoral Fellow
                       Institute for Catalysis, Hokkaido University

4. Publications

4.1. Papers

1. Kobayashi H, Yabushita M, Komanoya T, Hara K, Fujita I, Fukuoka A (2013) High-Yielding One-Pot Synthesis of Glucose from Cellulose Using Simple Activated Carbons and Trace Hydrochloric Acid. ACS Catal 3 (4):581–587
2. Fukuoka A, Kobayashi H, Yabushita M (2013) High-efficient saccharification of non-food biomass by using activated carbons (Japanese title: Kasseitan wo mochi-ita hikasyoku biomass no koukouritsu touka). Clean Energy 22 (8):45–48
3. Yabushita M, Kobayashi H, Fukuoka A (2014) Catalytic transformation of cellulose into platform chemicals. Appl Catal B Environ 145:1–9
4. Kobayashi H, Yamakoshi Y, Hosaka Y, Yabushita M, Fukuoka A (2014) Production of sugar alcohols from real biomass by supported platinum catalyst. Catal Today 226:204–209
5. Yabushita M, Kobayashi H, Hasegawa J, Hara K, Fukuoka A (2014) Entropically Favored Adsorption of Cellulosic Molecules onto Carbon Materials through Hydrophobic Functionalities. Chem Sus Chem 7 (5):1443–1450
6. Yabushita M, Kobayashi H, Hara K, Fukuoka A (2014) Quantitative evaluation of ball-milling effect on hydrolysis of cellulose catalysed by activated carbons. Catal Sci Technol 4 (8):2312–2317
7. Shi Y, Nabae Y, Hayakawa T, Kobayashi H, Yabushita M, Fukuoka A, Kakimoto M (2014) Synthesis and characterization of hyperbranched aromatic poly(etherketone)s functionalized with carboxylic acid terminal groups. Polym J 46 (10):722–727

8. Yabushita M, Kobayashi H, Shrotri A, Hara K, Ito S, Fukuoka A (2015) Sulfuric Acid-Catalyzed Dehydration of Sorbitol: Mechanistic Study on Preferential Formation of 1,4-Sorbitan. Bull Chem Soc Jpn 88 (7):996–1002

9. Kobayashi H, Yabushita M, Hasegawa J, Fukuoka A (2015) Synergy of Vicinal Oxygenated Groups of Catalysts for Hydrolysis of Cellulosic Molecules. J Phys Chem C 119 (36):20993–20999

10. Chung P-W, Yabushita M, To AT, Bae YJ, Jankolovitz J, Kobayashi H, Fukuoka A, Katz A (2015) Long-Chain Glucan Adsorption and Depolymerization in Zeolite-Templated Carbon Catalysts. ACS Catal 5 (11): 6422–6425

11. Yabushita M, Kobayashi H, Kuroki K, Ito S, Fukuoka A (2015) Catalytic Depolymerization of Chitin with Retention of N-Acetyl Group. Chem Sus Chem 8 (22):3760–3763

4.2. Book

1. Kobayashi H, Yabushita M, Fukuoka A (2016) Depolymerization of Cellulosic Biomass Catalyzed by Activated Carbons. In: Schlaf M, Zhang ZC (eds) Reaction Pathways and Mechanisms in Thermocatalytic Biomass Conversion I: Cellulose Structure, Depolymerization and Conversion by Heterogeneous Catalysts. Springer, Singapore, pp 15–26.

5. Patents

1. Fujita I, Fukuoka A, Kobayashi H, Yabushita M (2014) Method for decomposing plant biomass, and method for producing glucose. WO Patent 2014007295.

2. Fujita I, Yoneda T, Fukuoka A, Kobayashi H, Yabushita M (2014) Plant-biomass hydrolysis method. WO Patent 2014097799.

3. Fujita I, Yoneda T, Fukuoka A, Kobayashi H, Yabushita M (2014) Plant-biomass hydrolysis method. WO Patent 2014097800.

4. Fujita I, Yoneda T, Fukuoka A, Kobayashi H, Yabushita M (2014) Plant-biomass hydrolysis method. WO Patent 2014097801.

6. Awards

1. Nitobe Prize
Hokkaido University, Jun. 2008

2. Best Presentation Award at Winter Meeting of the Hokkaido Branches of Chemical Societies in 2011
Hokkaido Branch of Catalysis Society of Japan, Mar. 2011

3. Clark Award
   Hokkaido University, Mar. 2011
4. Best Poster Award at 52th Aurora Seminar
   Hokkaido Branch of Catalysis Society of Japan, Jul. 2011
5. Best Poster Award at 111th Catalysis Society of Japan Meeting
   Catalysis Society of Japan, Mar. 2013
6. Chinese Journal of Catalysis Best Poster Award at 2nd International Congress
   on Catalysis for Biorefineries
   Chinese Journal of Catalysis, Sep. 2013
7. Springer Theses Prize
   Springer Science+Business Media, Aug. 2015

Printed in the United States
By Bookmasters